图书在版编目 (CIP) 数据

美国建筑竞赛 . 1 / 美国《竞赛》杂志编辑部编；
常文心译 . — 沈阳 : 辽宁科学技术出版社 , 2017.6
　　ISBN 978-7-5591-0137-2

　　Ⅰ . ①美⋯ Ⅱ . ①美⋯ ②常⋯ Ⅲ . ①建筑设计 –
作品集 – 世界 – 现代 Ⅳ . ① TU206

中国版本图书馆 CIP 数据核字 (2017) 第 072538 号

出版发行：辽宁科学技术出版社
　　　　　（地址：沈阳市和平区十一纬路 25 号 邮编：110003 ）
印　刷　者：辽宁新华印务有限公司
经　销　者：各地新华书店
幅面尺寸：225mm × 295mm
印　　张：18.75
插　　页：4
字　　数：200 千字
出版时间：2017 年 6 月第 1 版
印刷时间：2017 年 6 月第 1 次印刷
责任编辑：杜丙旭
封面设计：李　莹
版式设计：李　宁
责任校对：周　文

书　　号：ISBN 978-7-5591-0137-2
定　　价：298.00 元

编辑电话：024-23280360
邮购热线：024-23284502
E-mail: 1207014086@qq.com
http://www.lnkj.com.cn

AMERICAN ARCHITECTURE COMPETITIONS 1
美国建筑竞赛 1

美国《竞赛》杂志编辑部 编
常文心 译

辽宁科学技术出版社
沈阳

CONTENTS

目 录

6　　**Introduction**
　　　前言

8　　 **Darat King Abdullah Ⅱ Arts Center Competition**
　　　达拉特国王阿卜杜拉 Ⅱ 艺术中心竞赛

12　　Zaha Hadid Architects
16　　Delugan Meissl Associates Architects
20　　Snøhetta AS
24　　Other contestants' schemes

32　　**Estonian National Museum Competition**
　　　爱沙尼亚国家博物馆竞赛

36　　Dan Dorell, Lina Ghotmeh, Tsuyoshi Tane
38　　ALA Architects
40　　Bramberger Architects – Atelier Thomas Pucher
42　　Other contestants' Schemes

44　　**The Gyeonggi-do Jeongok Prehistory Museum Competition**
　　　京畿道前谷史前博物馆竞赛

48　　 X-TU Architects
52　　Paul Preissner
58　　Easton+Combs

60　　**Mammoth and Permafrost Museum Competition**
　　　猛犸象和冻土博物馆竞赛

64　　Soren Robert Lund Arkitekter

66　　**Michigan State University Art Museum Competition**
　　　密歇根州大学艺术博物馆竞赛

70　　Zaha Hadid Architects
74　　Coop Himmelblau

86　　**Museum for L' Universitiare catholique de Louvain Competition**
　　　比利时勒芬天主教大学新博物馆竞赛

90　　Perkins+Will/ Emile Verhaegen
94　　Fuksas Architects
100　　TECTONICS ARCHITECTS LTD.
104　　Charles Vandenhove et Associes

106 **Nam June Paik Museum Competition**
白南准博物馆竞赛

110 Kirsten Schemel

112 Kyu Sung Woo

114 Noriaki Okabe

118 Other contestants' schemes

120 **San Jose State University Art Gallery Competition**
圣荷西州立大学美术馆竞赛

124 WW

128 Other contestants' schemes

134 **The Perm Contemporary Art Museum Competition**
彼尔姆当代艺术博物馆竞赛

138 Bernaskoni

142 Valerio Olgiati

144 Zaha Hadid Architects

150 Acconci Studio+Guy Nordenson and Associates

154 Alexander Brodsky Bureau

156 Asymptote Architecture PLLC

158 Esa Ruskeepaa

160 A-B

164 Søren Robert Lund Arkitekter

168 Totan kuzembaev Architecture Workshop

170 **University of Michigan Museum Addition Comprtition**
密歇根大学博物馆附楼竞赛

174 Allied Works Architecture

180 The Polshek Partnership

182 Weiss/Manfredi Architects

186 **Whatcom Museum of History & Art and Whatcom Children's Museum Competition**
霍特科姆艺术历史博物馆和儿童博物馆竞赛

190 Cambridge Seven Associates, Lnc./Donnally Architects

194 **Czech National Library Competition**
捷克国家图书馆竞赛

198 Duture Systerms

206 Carmody Groarke

212 HSH architeki

220 EMERGENT

234 John Reed

242 MVMarchitekt + Starkearchitektur

244 Dagmar Richter

248 Holzer Kober Architekturen

252 **Kazakhstan Library Competition**
哈萨克斯坦图书馆竞赛

256 BIG

258 **Milford Library Competition**
米尔福德图书馆竞赛

262 Frederc Schwartz Architects

268 Bohlin Cywinski Jackson

270 BKSK

272 **Philadelphia's Public Library Competition**
费城公共图书馆竞赛

276 Moshe Ssfdie & Associates

280 TEN Arquitectos

288 **Stockholm City Library Competition**
斯德哥尔摩城市图书馆竞赛

292 Heike Hanada

296 Other contestants' schemes

298 **Index**
索引

Darat King Abdullah Ⅱ Arts Center Competition

达拉特国王阿卜杜拉Ⅱ
艺术中心竞赛

竞赛背景　Competition Background

Located at the bottom of Amman's main valley and adjacent to Ali Bin Abi Talib road — the main traffic artery running through the valley — it is where the more affluent can almost rub elbows with the have-nots. With these social contrasts in mind, King Abdullah visualized an arts center accessible to all citizens regardless of social stature. To arrive at a design for the facility, the government engaged the German firm, [phase eins] led by Benjamin Hossbach, to administer a limited design competition. Thirty firms answered the call to submit expressions of interest.

The teams were asked to design a building consisting of a large theater accommodating 1600 persons and a small theater with 400 seats available, both equipped with highly sophisticated audio systems. Training areas and public facilities, such as a restaurant and a café will also be provided. The development requires a certain level of flexibility in terms of its design, construction and operation to accommodate large events and experimental workshops.

约旦首都安曼的艺术中心位于主要山谷的底部，毗邻阿里本艾比利布公路——贯穿山谷的主要交通要道——是安曼最富裕和最贫穷的人互相接触交往的地方。考虑到这些社会差异，国王阿卜杜拉想要建成一个艺术中心，不分社会地位，对所有居民开放。为此，政府委托本杰明·胡思巴赫领导的德国"phase eins"公司管理一次限制设计竞赛，30家建筑公司有意参赛。

竞赛要求设计团队设计一个由一间容纳1600人的大剧场和一个有400个坐席的小剧场组成的建筑物，两个剧场都要配备高端的音频系统，还要提供培训领域和诸如餐馆和咖啡厅的公共设施。开发要求在满足大型活动和体验车间要求的设计、施工和运行方面有一定程度的灵活性。

The sponsor anticipated that the competition would lead to a world class design and enable "Darat King Abdullah II" to serve as a venue for local community events as well as the cultural hub for artistic programs and activities.

竞赛赞助商预测竞争将会产生一个世界级的设计，使"达拉特国王阿卜杜拉 II"艺术博物馆成为一个当地社区活动场所和一个艺术项目和活动的文化中心。

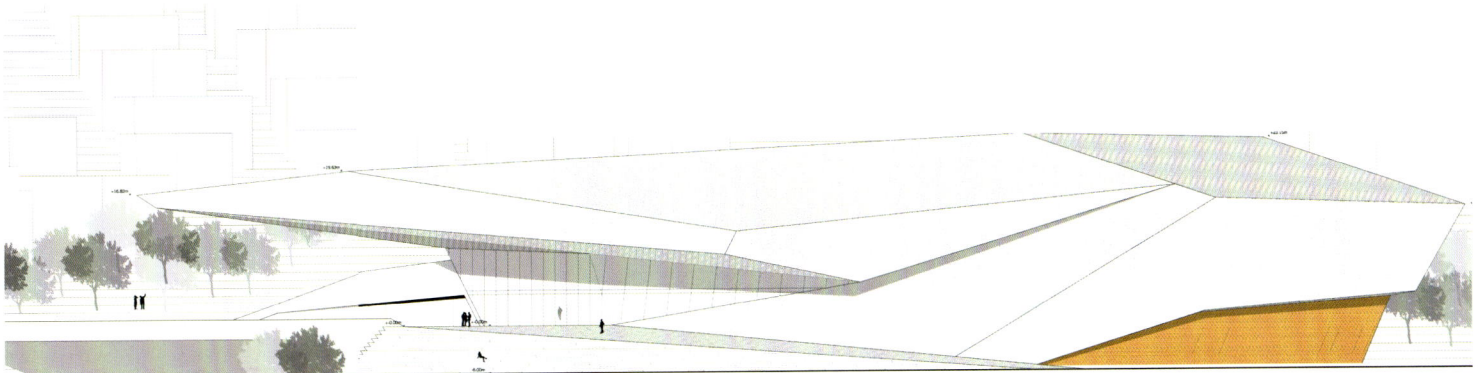

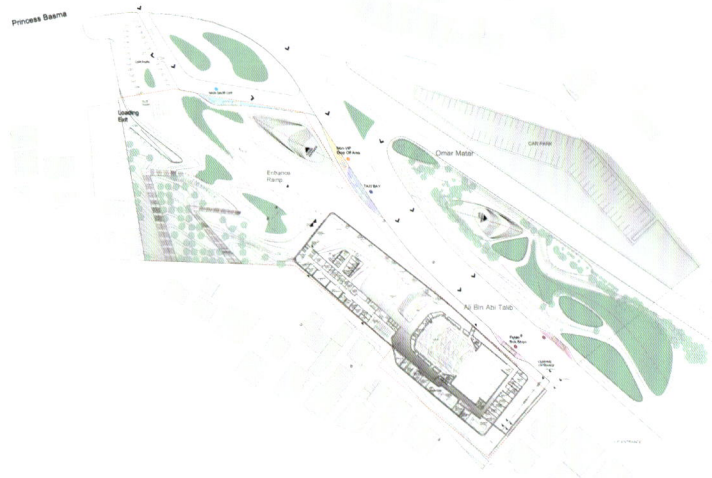

扎哈·哈迪德建筑事务所
参赛项目1（一等奖）

Zaha Hadid Architects Project 1
the first prize

The jury obviously was impressed by the poetic nature as well as the siting of the Hadid design: Beyond the first sight attractions, a closer examination of the proposal started to reveal the special attributes regarding the approach to the site and the context, the suggested weightlessness of the opaque monolith and the unavoidably felt, massive presence of the emptiness. The shifting interceptions of light by the hyper-spatial surfaces evoked the timelessness of wind worn memories of the mountains of the region. One could hear the wind in, through and around the cave like interfaces among the building and the site so generously liberated and enhanced through pushing the building to the Northeast edge of the site.

Otherwise, concern was voiced about the excessive ratios between the gross floor area and the utilization area, the challenges of large cantilevers and the very tall glass enclosures, especially in an earthquake zone. Also, they criticized the lack of attention paid to the small theater: the technical consultants felt it did not fulfill all of the technical requirements. Finally, the consultants felt that the volume of the building would have to be reduced to meet budget requirements.

评审委员会显然非常喜欢扎哈·哈迪德建筑事务所设计的诗意和选址：如果看完方案第一眼之后再进行仔细观察，就会发现方案在项目地点和环境背景的处理方法、不透明整块石料的无重状态、大量存在的不可避免的空虚感等方面的特殊特点。通过多维空间表面截取的不断变化的光线激发起人们对这一地区山脉风蚀的永恒记忆。人们在建筑物和项目地点内洞穴状接口处可以听见风吹进来、吹过和吹绕的声音，如此之多，如此之强，把建筑推到项目地点的东北边缘。

另一方面，评审委员会也就建筑总面积和使用面积之间的过度比值，尤其是在地震灾区出现的大悬臂和高玻璃外壳的挑战等问题表示出疑虑。评审委员会也批评了方案没有注意到小剧场的要求；技术顾问认为方案没有满足项目所有的技术要求。最后，顾问们觉得建筑体积将必须被减少才能满足预算要求。

Daylight perspective 日间景观

Main theater 主剧场

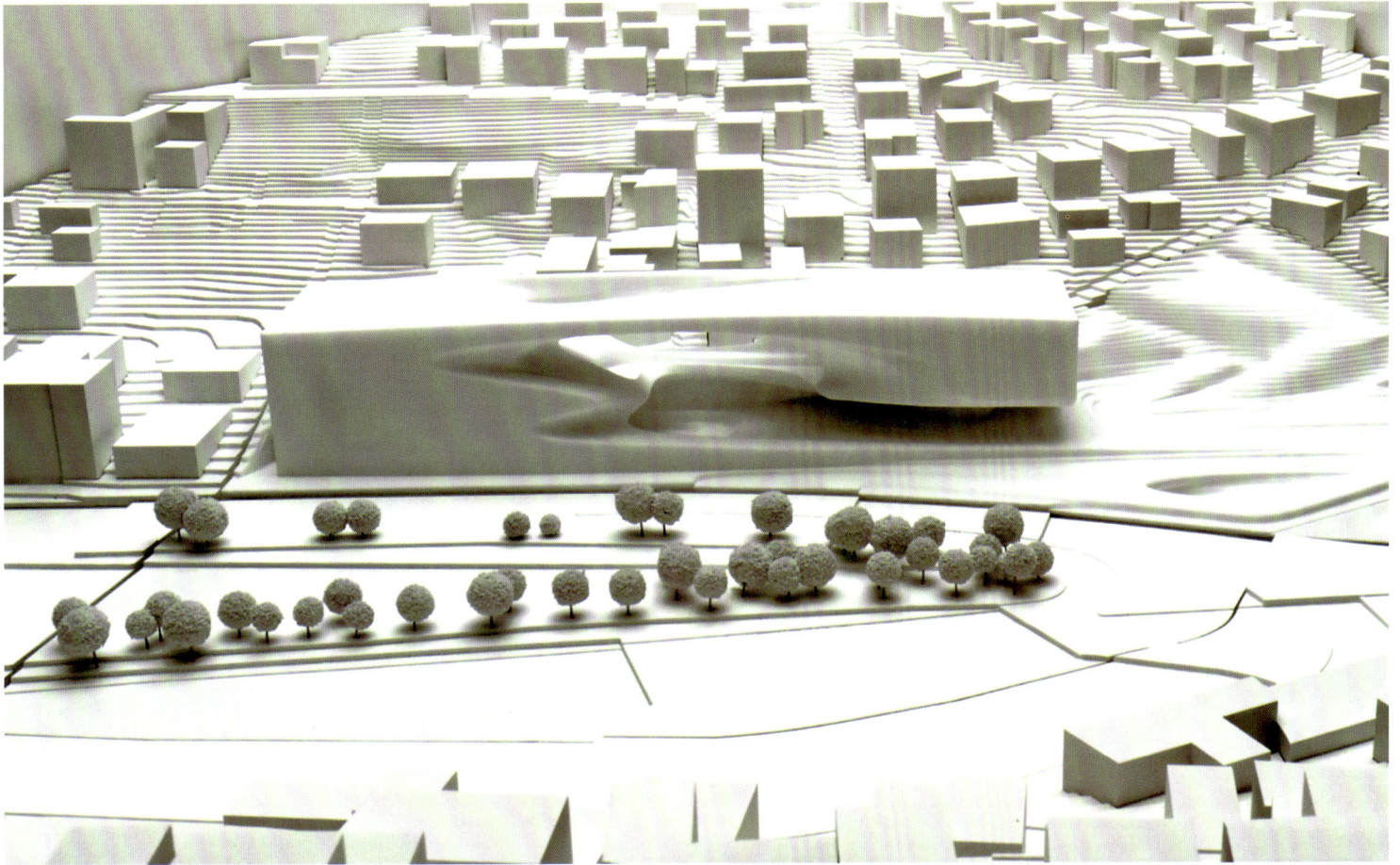

德卢甘·梅斯建筑事务所
参赛项目2（二等奖）

Delugan Meissl Associates Architects
Project 2 the second prize

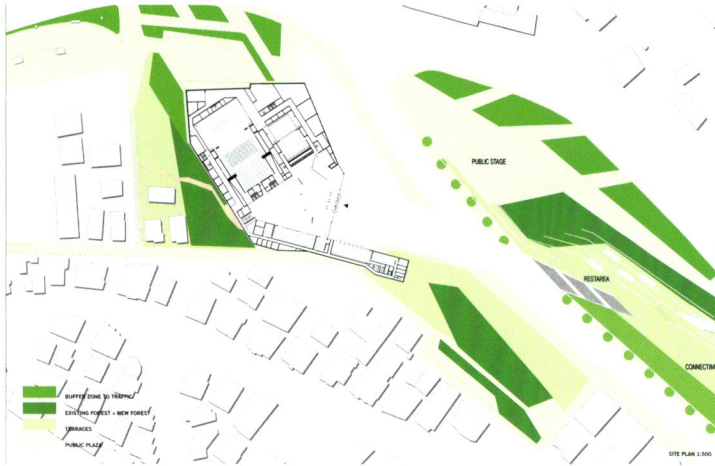

SITE PLAN 1:500

The jury praised the organization, circulation and general integration of the spaces to be the strong point of Delugan Meissl Associates Architects' design. They expressed their approval of the sculpturally dramatic exterior, which they felt could be easily executed. The functionality of the theaters was questioned, particularly the large concert hall, which was in the traditional shoe-box shape. Both theaters should be reconfigured so as to accommodate more types of events. Also, a couple of elevations were overpowering and needed to be reduced. But they were very impressed with the "powerful formal statement" of the design.

评审委员会赞扬了德卢甘·梅斯建筑事务所设计方案的组织、流通和空间整合，认为这些都是该设计的优点。他们同意方案刻纹装饰戏剧性的外观，觉得这样容易施工。但是评审委员会就影院的功能，特别是对传统包厢形状的大音乐厅的功能提出质疑。两座剧场都应该重新配置，以满足更多类型活动的需要。此外，两个立视图气势过强，需要减弱。但是评审委员会对于该方案的"强有力的形式表达"印象十分深刻。

View to foyer 大厅

Aerial view at night 夜间鸟瞰图

Large theater 大剧场

斯诺赫塔建筑事务所 参赛项目3（三等奖）

Snøhetta AS Project 3
the third prize

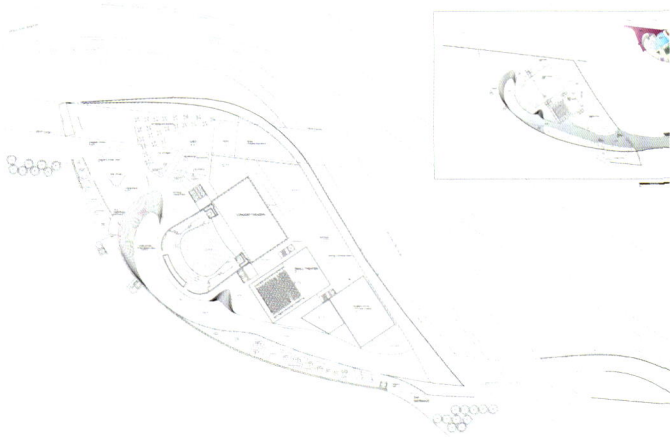

The jury liked the "clarity of the project and its effective functional organization" of the Snøhetta design. It was considered a "strong and poetic statement". The public circulation and dramatic programmatic wall that borders the spine was regarded as one of the best features of the design. However, the jury expressed reservations about the relation between the roof structure and the inner constructions as well as its feasibility within the budget. Also the entrances to the building were criticized for "lacking generosity and attraction". But the jury still felt that, with changes, it could be developed into a feasible project.

评审委员会喜欢斯诺赫塔建筑事务所设计方案的清晰性和组织结构的功能有效性，认为该项目是一个"强大和诗意表达"。公众流通和沿脊柱结构分布的戏剧性墙体被视为该设计最佳的特点之一。然而，评审委员会表达了对"屋顶结构和内部结构之间关系，以及预算之内的可行性"问题的保留意见。此外，建筑入口因其"缺乏大气和吸引力"受到批评。但是评审委员会仍认为，该方案在改变之后可以成为一个可行性方案。

Elevation 立面图

Interior view 室内

Night aerial view 夜间鸟瞰图

其他参赛方案 **Other contestants' schemes**

SECTION A 1:200

Section 剖面图

1:200 SECTION B

Section 剖面图

Vignettes 装饰花纹

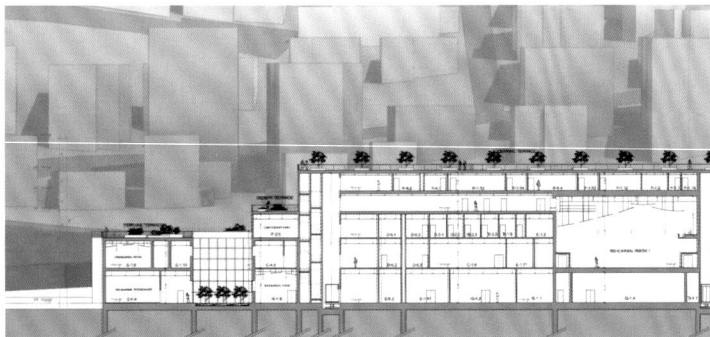

Transversal section 横向剖面图

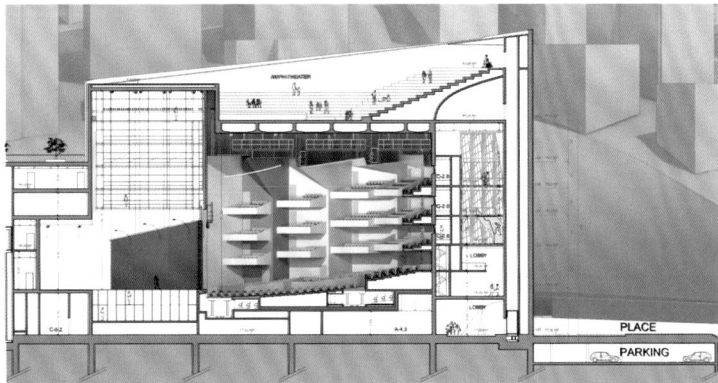

Longitudinal section 纵向剖面图

Night view 夜景图

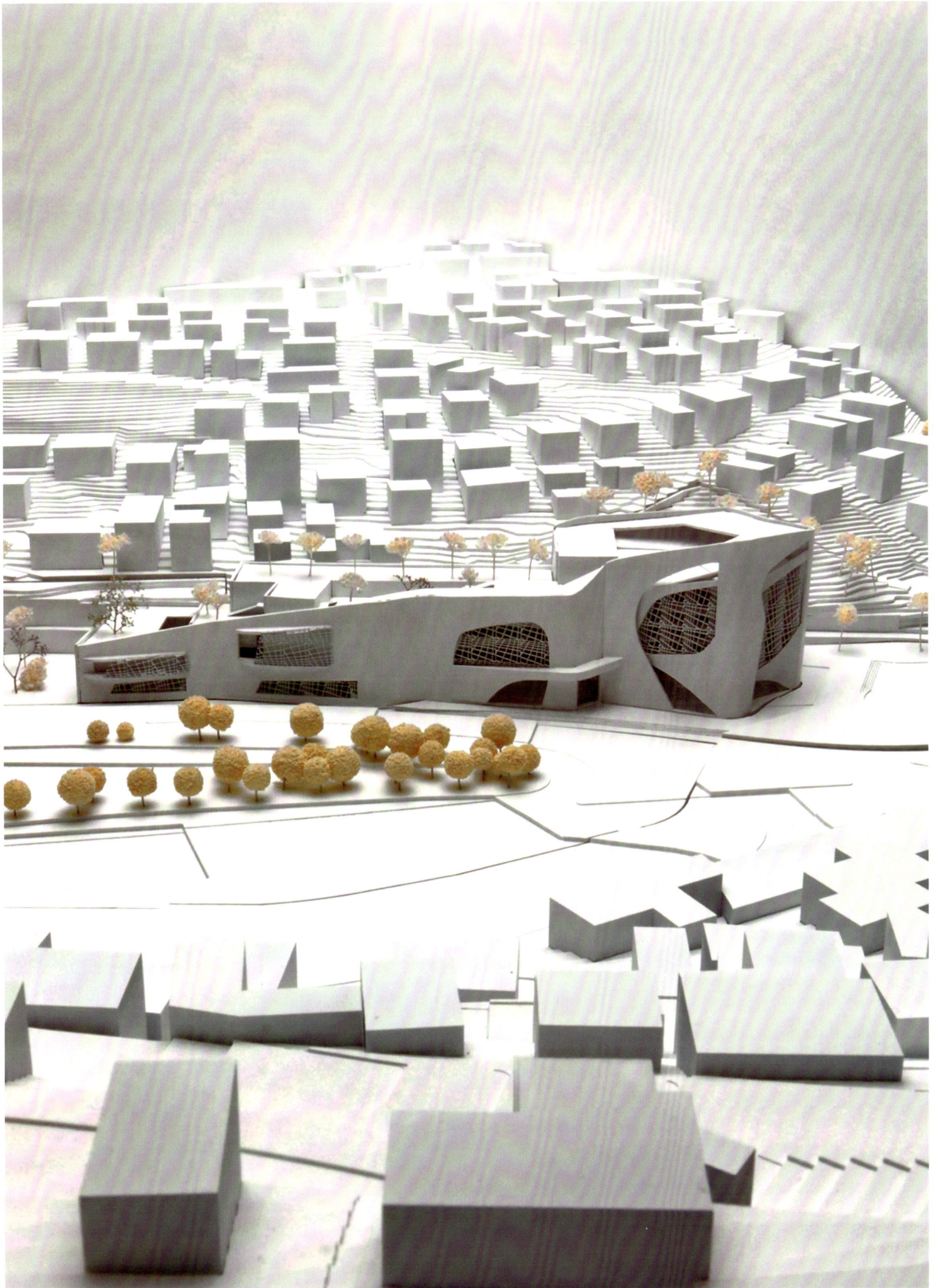

Aerial view 鸟瞰图

longitudinal section 1:200

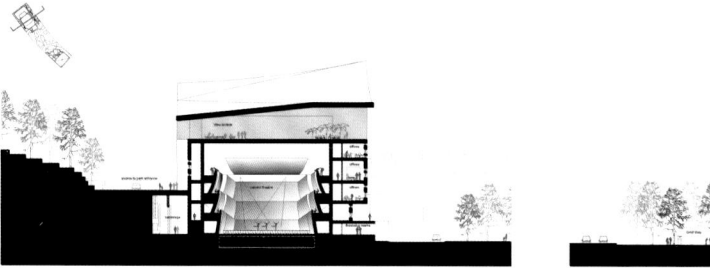

section through concert theater 1:200

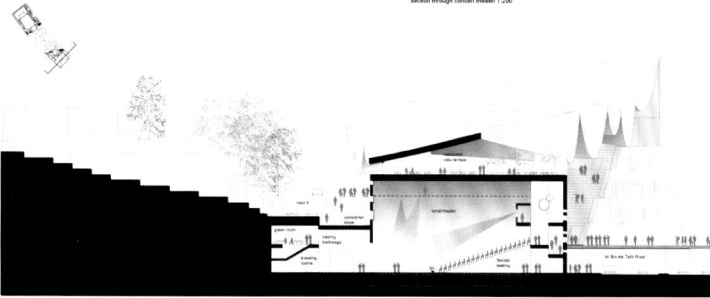

Section 剖面图

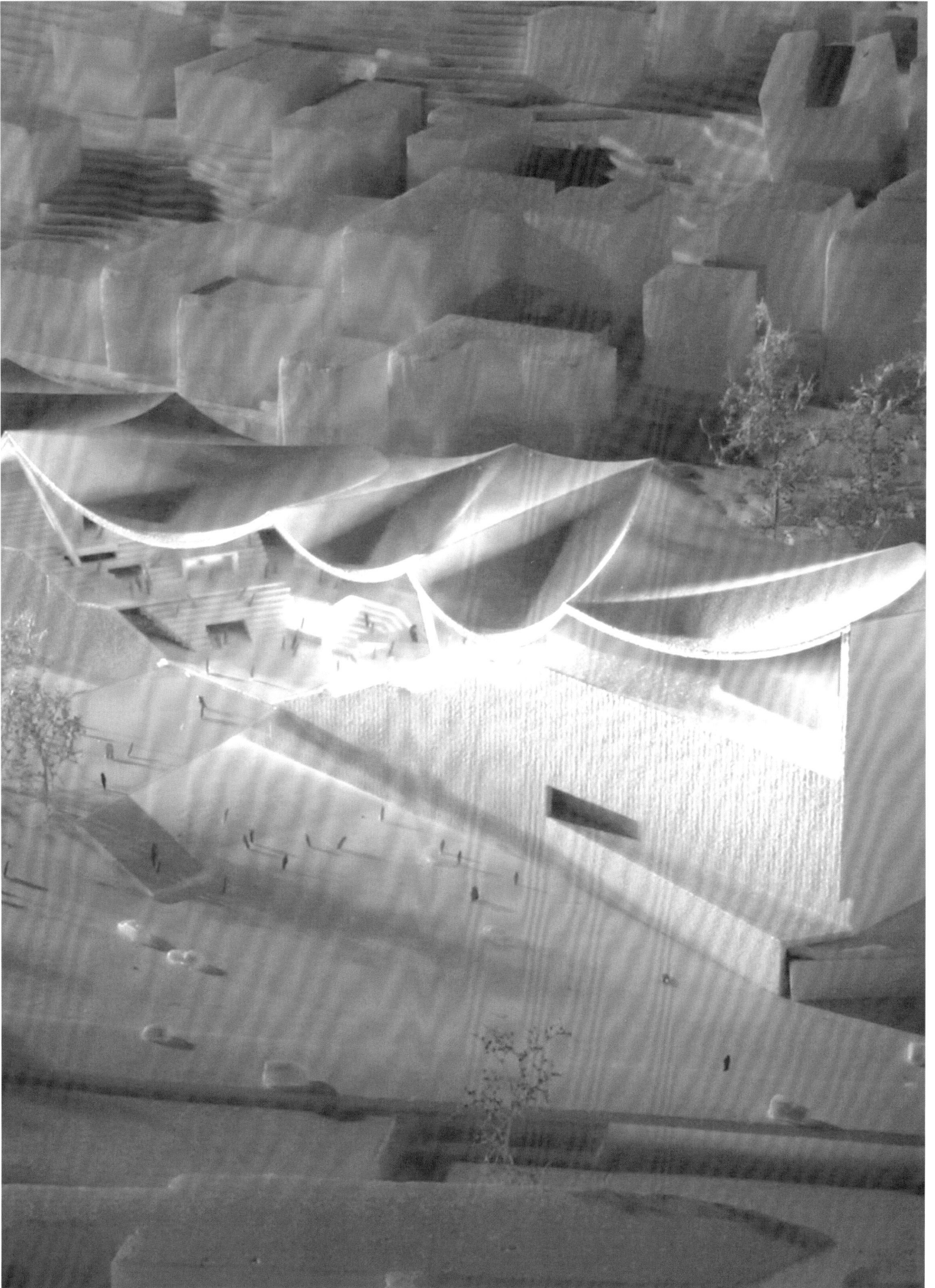

Aerial view 鸟瞰图

Estonian National Museum Competition

爱沙尼亚国家博物馆竞赛

Competition Background

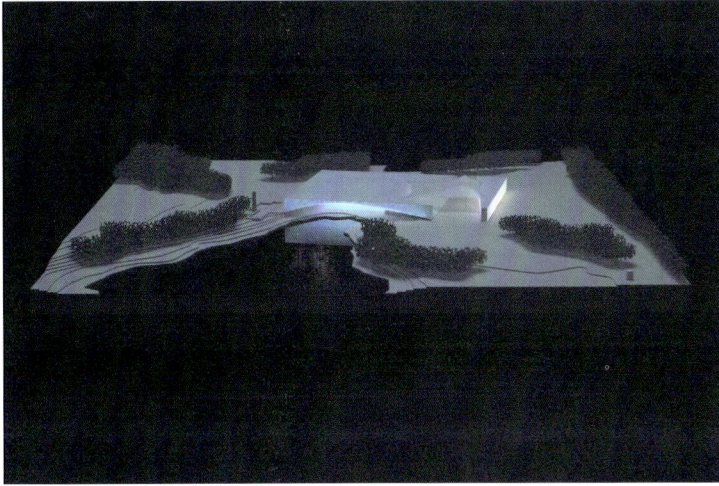

Architecture in the service of the state can be a powerful tool. It lets a faceless bureaucracy present itself as a real thing. It can make ideals and memories into concrete facts. The competition for a new National Museum in the small country of Estonia shows the potential and pitfalls of such a national architecture. While it can define the state, what does that mean when the state as a concept is a difficult idea these days? What future can Estonia imagine for itself? What is there left for architecture to represent?

The answer, if we can believe the results of the recent Estonian National Museum Competition, is that architecture can use place above all else for meaning. Everything human beings have done with a place, everything they have built there, and every association they have with a site as part of a much larger whole is the basic material the architect can use to design a construction that will bring out all of this history and all of these latent associations.

建筑是为国家服务的强有力工具之一。它让没有实体感的政府变成实体，把理念和回忆凝成宏大的事实。爱沙尼亚国家博物馆竞赛正是体现了国家级建筑的潜力和缺陷。尽管建筑能定义一个国家，但是如今国家概念已经越来越难定义了。作为一个小国，爱沙尼亚国家的未来是怎样的？又怎样用建筑来呈现？

如果我们可以相信爱沙尼亚博物馆竞赛的结果的话，答案就是建筑可以创造高于一切的意义。每块人类建设过的土地，每栋人类建造的房子，每段人类和土地的关系，都是建筑师可以运用的材料，他们用这样的建筑展示历史和潜在的联系。

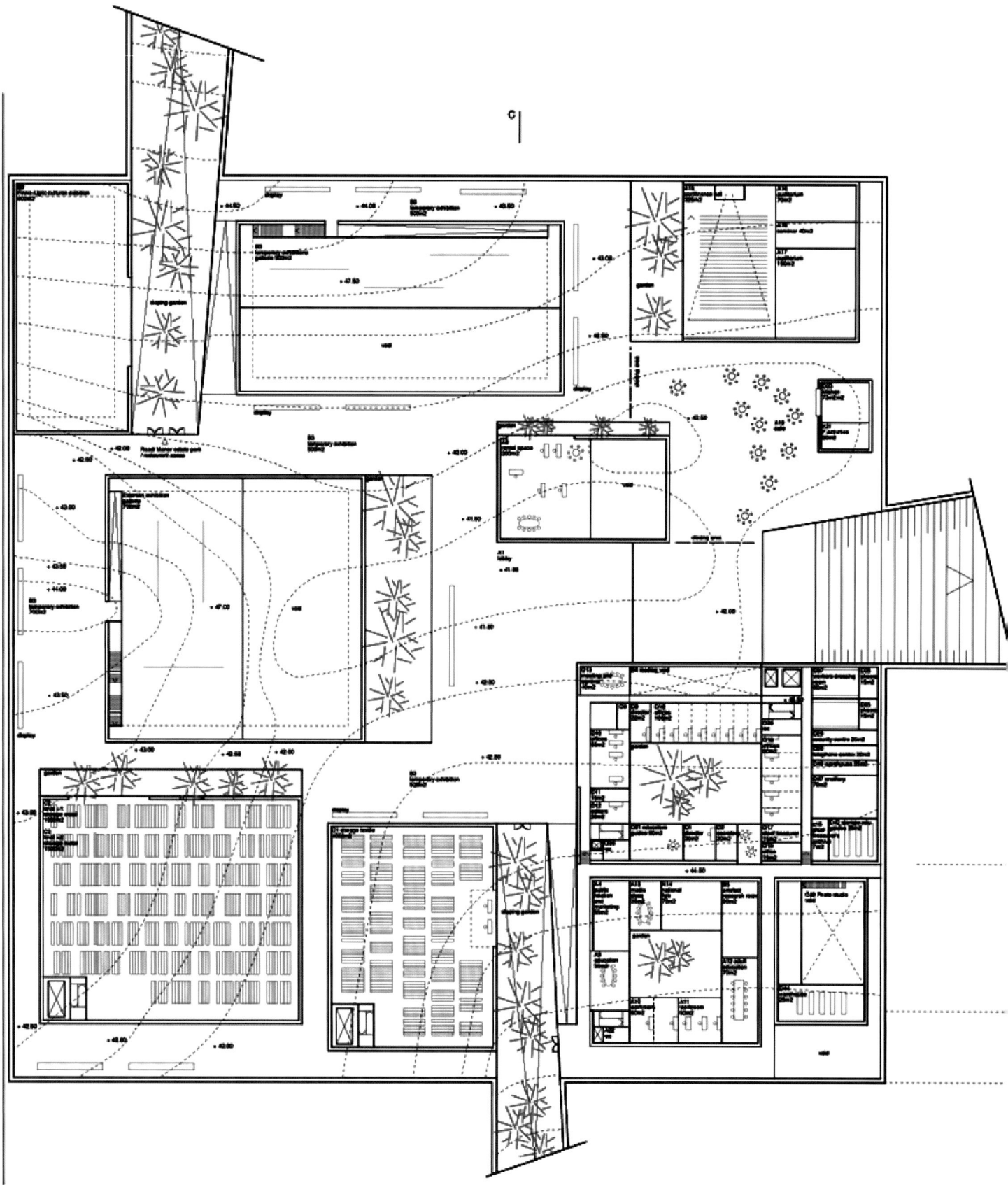

The "Memory Field" is a design that really makes them part of Europe in its design," Miny Maas says, "Not just because of the multi-national design team, but also because the design is part of an international attempt to create a modern architecture that can be monumental and mark the landscape. But it is also so right for this site, it makes it seem beautiful." Its command of the site, its romantic attitude towards the ruins of the military-industrial complex, and its grand spaces will all serve to give Estonian National Museum a setting appropriate to its collections and its site.

评委米尼·麦阿斯称:"获得一等奖的'记忆之地'的设计真正融入了欧洲设计。并不仅仅是因为它的设计团队是国际化的,也因为这个设计尝试创造一个具有纪念意义并且能标志风景的现代建筑,这是一个国际性的尝试。它和周围的风景是那么融合,让风景看来更美丽。"
"记忆之地"将是爱沙尼亚文化和建筑浓墨重彩的一笔。它和风景的融合度,它对于军工厂遗址浪漫的态度,它宏大的空间,都会让爱沙尼亚国家博物馆与它的藏品和景观相称。

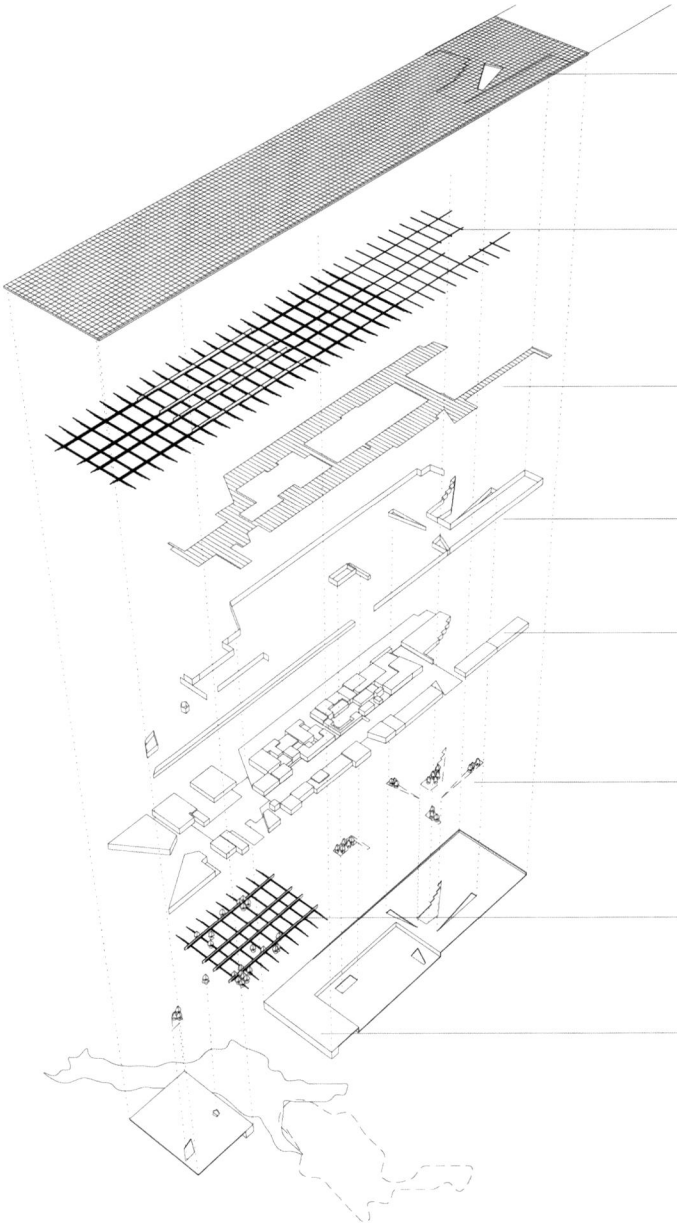

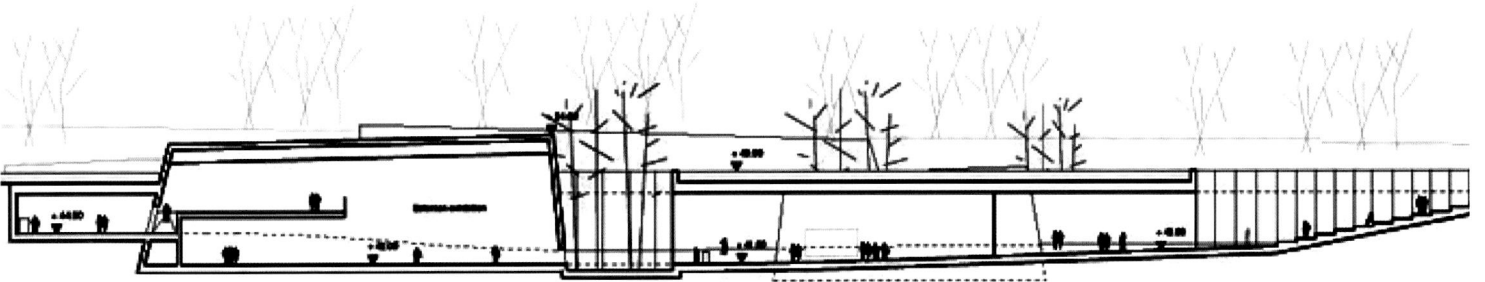

丹·多莱尔、丽娜·勾特曼、和朱由希·泰恩
参赛项目1（一等奖）

Dan Dorell, Lina Ghotmeh, Tsuyoshi Tane
Project 1 the first prize

The winning entry — "Memory Field" emphasizes its location by continuing the runway's line literally across the tip of the lake, forming a bridge containing the major program elements. Visitors would move from a parking area to what is in the design proposal an immense cantilevered portico. After passing all the necessary public services and education spaces, they would go straight through to a central exhibition hall, buttressed to the South by museum offices and to the North by collection workshops and storage spaces. Temporary exhibitions would occupy a parallel gallery on the building's North side, where floor-to-ceiling glass would provide a view over the lake. After passing through the central exhibitions, visitors would be able to keep walking, ascend broad steps and find themselves on the roof looking out down the axis of the runway and over the flat terrain. Though the designers imagined sculptures up on the roof, its main function is clearly to serve as a "dramatic space," the jury noted. The jury feel that the scheme's clear idea could be developed into a building that would be both functional and symbolically clear.

The winning entry focuses almost completely on the site. The collections are placed in such a way to seem as if they are an interruption on the way to the roof. In the extremely evocative renderings the winning team submitted, one only sees abstract sculpture, such as a Louis Bourgeois "Spider," occupying what appear to be vast spaces. The historical content appears to be secondary to the scheme. This is a museum that could be, strangely enough, anywhere. The same can be said for most of the entries the jury selected for prizes and mentions. They are all more or less isolated objects sitting in the landscape, with a clear sculptural presence but little sign of thought give to that mythical "Finno-Ugric" culture.

"记忆之地"将坡道延伸到了湖边，让主建筑变成连接二者的桥梁，从而突出了项目的地理条件。参观者可以穿过停车区域到达一个巨大的悬臂式门廊。在参观完所有的公共服务设施和教育展馆后，他们可以一直走到中心展厅。中心展厅拱壁的南端是博物馆办公室，北端是藏品工作室和仓库。临时的展出可以安排在主建筑北侧的平行展厅，透过那里的落地玻璃窗可以观看湖景。穿过中心展厅，参观者可以继续前行，登上楼梯，到达屋顶，俯瞰坡道和平坦地带。尽管设计师也想在屋顶上安置一些雕塑，但是评委们称屋顶的主要功用是作为一个"有戏剧性效果的空间"。评审委员会认为这一设计清晰的理念足以建成一座仅实用又简洁的建筑。

"记忆之地"的设计重点遍布整个场地。屋顶藏品的摆放方式似乎对屋顶的设计造成了干扰。在一幅让人印象深刻的效果图里，只有一座抽象的雕塑占据了整个庞大的空间。在这个设计中，历史性似乎已经屈居于第二位。由于缺乏地域性，这是一座可以建在任何地点的博物馆。这也是所有获奖和入围项目的通病。这些项目都多多少少的脱离了周围的环境，形象独特却没能表现出神秘的乌戈尔族文化。

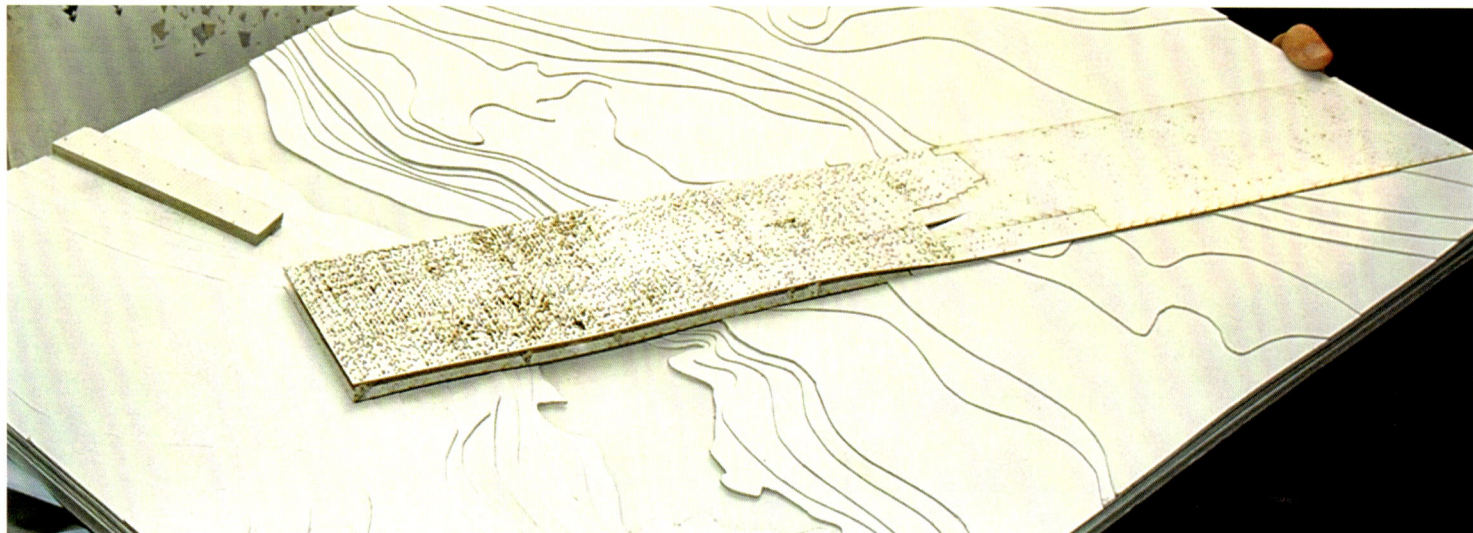

ALA建筑事务所 参赛项目2（二等奖）

ALA Architects Project 2
the second prize

Axonometric view 轴测图

The Second Prize winner, a Finnish entry by the firm ALA, "Gems," seems a particularly ham-fisted block, its side curled up to symbolize something unnamed while providing entry to a rooftop auditorium. In a strange gesture, the offices are separated out into a bar-building that confronts visitors as they approach the building from the parking lot and that is connected to the main workspaces by a tunnel.

ALA公司设计的项目——"宝石"，外观像一块大砖头。建筑的一侧微微卷起，象征着未知的东西，同时也是通往屋顶礼堂的入口。奇怪的是，博物馆的办公楼是独立于博物馆之外的长条形小楼，和主工作室由地下通道相连，参观者一走出停车场就能看到。

Elevation 立面图

Exhibit space 展示空间

布莱姆伯格建筑事务所–托马斯·帕切工作室 参赛项目3（三等奖）

Bramberger Architects – Atelier Thomas Pucher
Project 3 the third prize

A much more evocative design is the Third Place winner, "Estonia is on the Verge of a New Millennium," designed by Bramberger Architects – Atelier Thomas Pucher. This is a cube covered in enameled glass cut into a pattern that the designers claim evokes the tradition of lacework.

Inside, the program elements are simply stacked, with public functions taking up the base, exhibition areas on the first, second and third floors, and non-public areas above that. The architects, however, did not solve the perennial problem in a vertical museum, which is how to achieve a graceful, inviting, and yet non-intrusive and efficient circulation method. Instead, obtrusive escalators, elevators and stairs stitch the floors together.

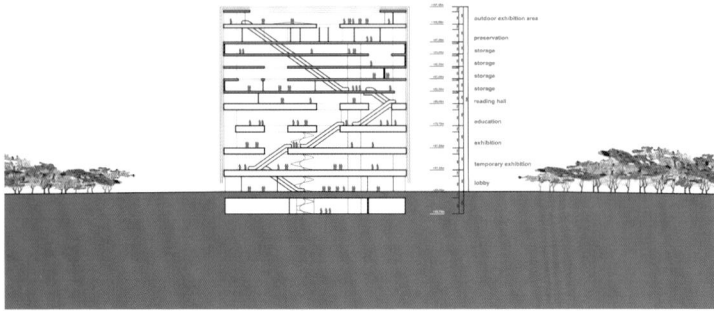

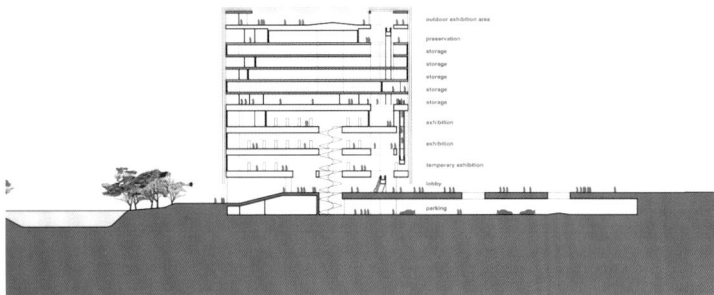

Section 剖面图

布莱姆伯格建筑事务所–托马斯·帕切工作室设计的"爱沙尼亚新纪元"是一个被釉面玻璃包裹住的立方体，玻璃雕刻形状的灵感来自于手工蕾丝制品。内部的结构是简单的堆叠：公共职能区域在底部，展区在一、二、三层，非公共区域在顶楼。但是建筑师并没有解决垂直空间的博物馆一直存在的问题——如何达成一个优雅、吸引人、不具干扰性而又有效的人流流通方式。相反的，他们使用障碍性的扶梯、电梯和楼梯将各个楼层连接在一起。

Exhibit spaces 展示空间

Exhibit spaces 展示空间

Exterior 外观

其他参赛方案 **Other contestants' schemes**

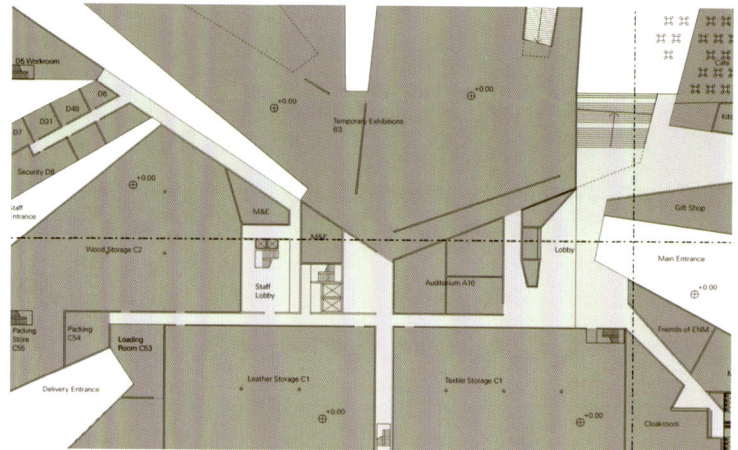

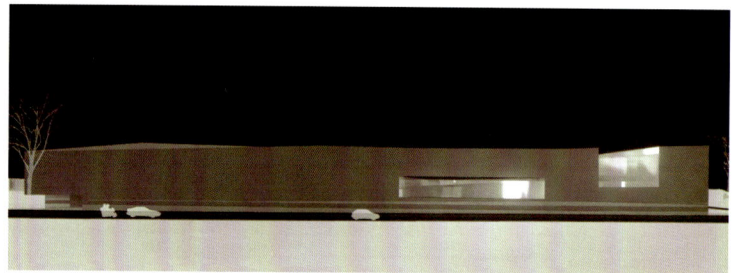

View from lake to museum 从湖边看博物馆

Perspective from lake 湖畔景观

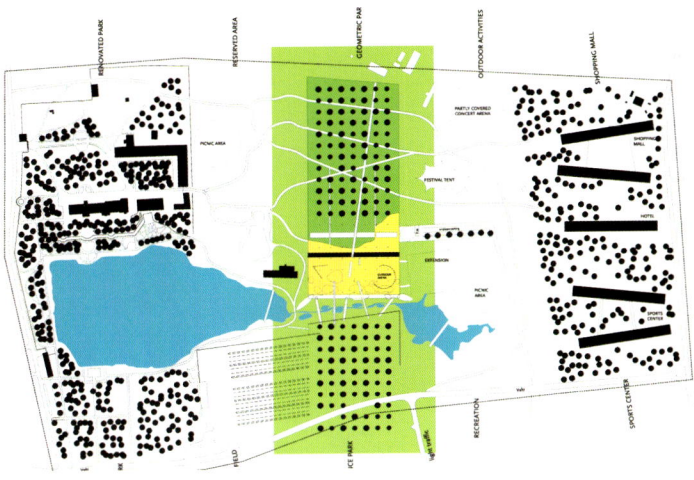

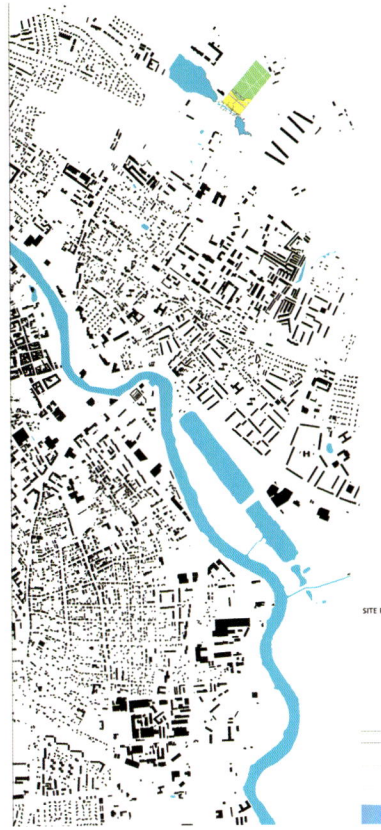

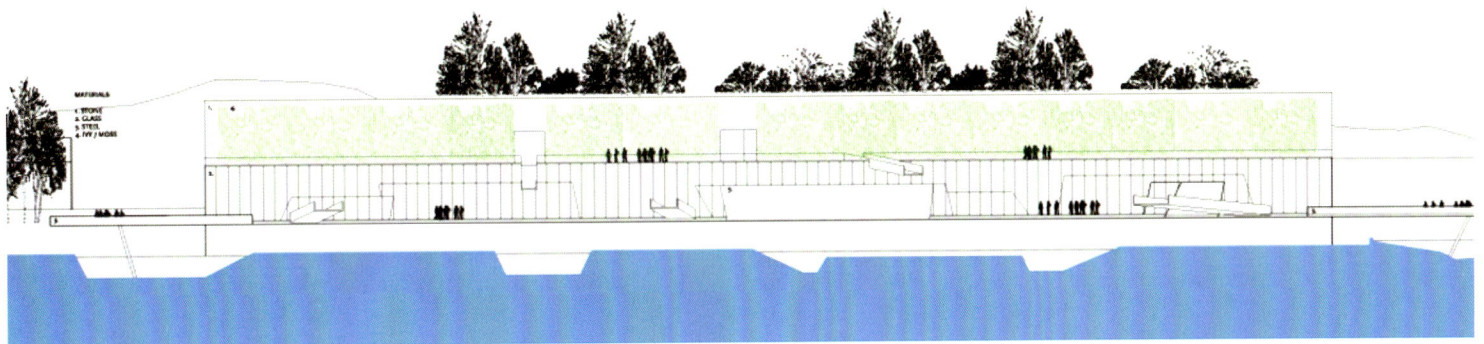

The Gyeonggi-do Jeongok
Prehistory Museum Competition

京畿道前谷史前博物馆竞赛

竞赛背景 Competition Background

Located near one of the world's most important paleolithic archeological sites, the design of the new Gyeonggi-do Jeongok Prehistory Museum presented a real challenge to competitors. The competition brief lacked specificity as to the museum's functionality—there were no hard and fast rules governing the organization of the display palette. Also, there was no Ralph Applebaum to choreograph circulation and exhibition spaces. The question was, "Is this more like a natural history and science museum?"

On top of this, the sponsor was interested in a design which would reflect cultural heritage. Still, one of the main criteria was to design a building which would "maximize the striking qualities of the excavation site." The participants were given a rough idea of the space requirements for each museum function, whereby at least one-third of the area was to be allotted to exhibitions.

建于世界上最重要的旧石器时代遗址，京畿道前谷史前博物馆的设计对所有参赛者都是一个挑战。评审委员会并没有对博物馆的功能结构有详细要求，设计师的设计不受约束，同时，竞赛对展示的空间大小也没有详细要求。重要的问题是"这个设计作品像不像一座历史科学博物馆"。

出于以上原因，340支参赛团队充分发挥自己的能力，进行设计。赞助商最关注的是设计能否反映文化底蕴，其次，他们要求设计可以将这片被挖掘的土地的优势最大化的展示出来。参赛团队的设计至少要有三分之一是展示空间。

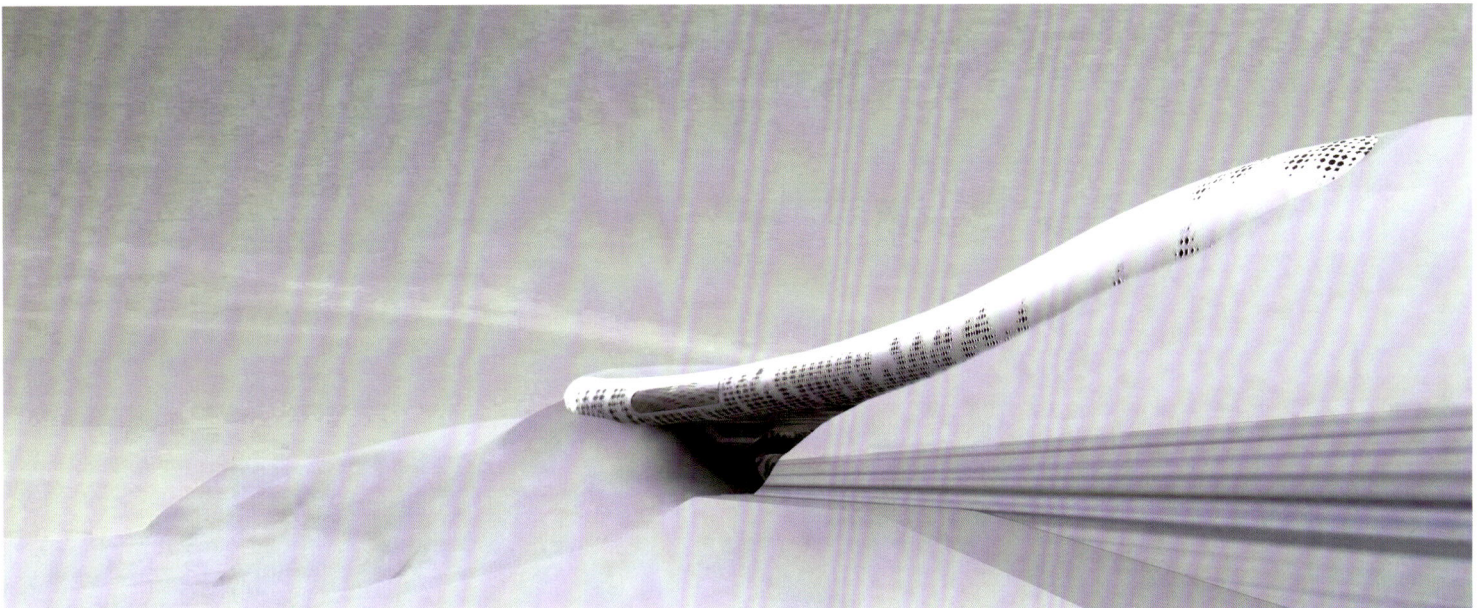

专家点评 **Remarks**

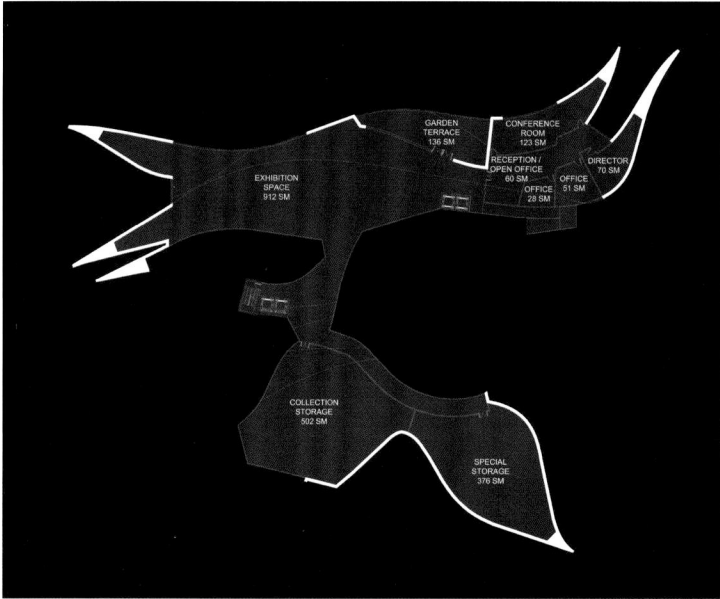

All the three designs are apt to the concept of prehistory museum and have bold but beautiful appearance. Comparatively speaking, X-TU's design gets along more harmoniously with the natural environment. Maybe that's why it is the winning design.

三个入围设计都十分符合史前博物馆的主题，并且三个设计的外观都十分大胆前卫。但是X–TU的设计与周围的环境最为和谐，这也许就是它获胜的原因之一。

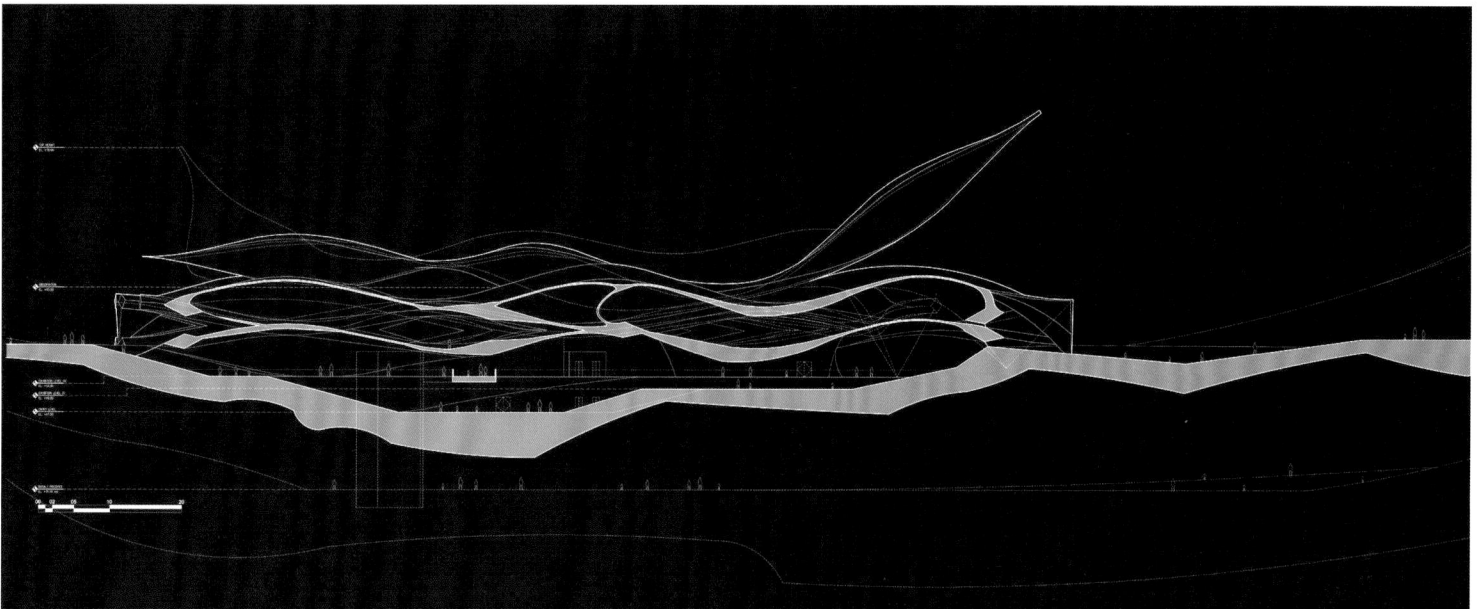

X-TU建筑师事务所 参赛项目1（一等奖）

X-TU Architects Project 1
the first prize

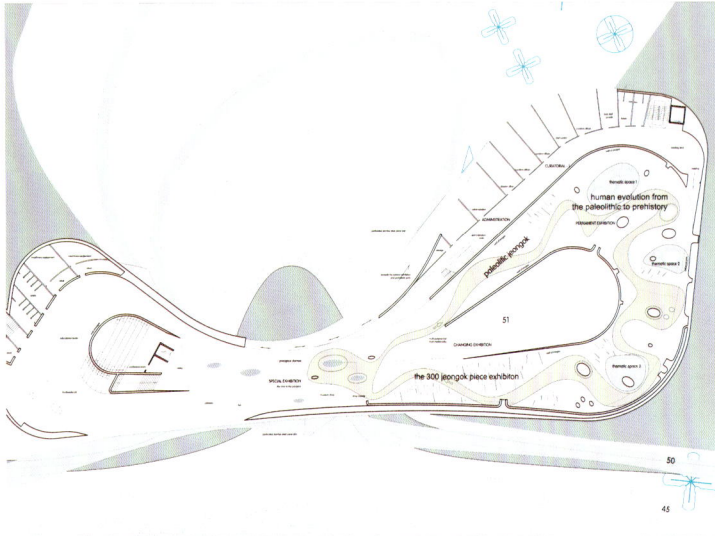

The design of the first place winner by X-TU Architects featured a shimmering skin which might suggest the appearance of a spaceship supported on two blufs, hovering over a depression in the landscape.

In the words of the winner, they "wished to honor the riverside landscape which saw the birth of the first inhabitants of Korea, and acknowledge the beauty of the two hill curves echoing the river meanders. How to enhance such a pre-existent form and its geological underground chasm? • by digging the chasm to let the Earth tell its history; • by alleviating the visual hold of the project in order to let the chasm express itself; • for this purpose, the building will be encased into the hill which has been hollowed out and the stockrooms wil be located underground; • by curving the central part of the building so as to unveil the geological crack (and also the sun, from the edge of the crack); • by clothing it in a 'shimmering skin' which will reflect the precipice from underneath.

Thus set up, the project appears like a bridge stretched over two hills which can be seen from a long distance, approaching from the motorway. The precipice as a natural threshold, and the emotion it induces will be used to realize a symbolic threshold into the 'prehistoric era' which will also give access to the 'prehistory park.' Then we create paths, many paths around the curves of the project and of the cliffs, because the paths which were made by the animals going down to the river to drink, belonged to the 'first human beings' landscape."

X-TU's treatment of the museum interior can only be characterized as symbolically organic in its language. The visitor is drawn from station to station by the existence of display modules resembling a moonscape, where it is possible to view what is transpiring below — bringing the viewer closer to the reality of the "dig."

X-TU建筑师事务所的获奖设计闪闪发光，像两块断壁支撑的宇宙飞船，盘旋于遗址地面之上，给人以萧索之感。

建筑师们称他们想要"向孕育了第一代朝鲜人的流域致敬，并且充分表现河畔的两座高山的美丽"。对于如何增添这早就存在的风景和地质断层的魅力，他们采取了以下的方法：挖掘断层，让大地展示历史；减轻项目外观的视觉性，以突出断层；将博物馆隐藏进被挖掘中空的山，并将储藏室安排在地下；让博物馆的中心呈弧线形，以展现地质断层（还有从断层边缘升起的红日）；给博物馆披上"闪闪发光的外衣"，以反照地下的断层。

完工后，博物馆像两山之间的一座桥，从很远的地方都可以看到。断层是一个天然的门槛，它所表达的情绪让人身临"史前时代"，是"史前公园"的入口。设计师还在博物馆和悬崖周围设计了许多小道，就像动物前往河边喝水时留下的轨迹一样，只不过现在它们属于人类。

X-TU的室内设计是有机式的。参观者一站一站的浏览，展台的矮柱模仿了月球表面，可以通过展台看到地下，让参观者更接近挖掘现场。

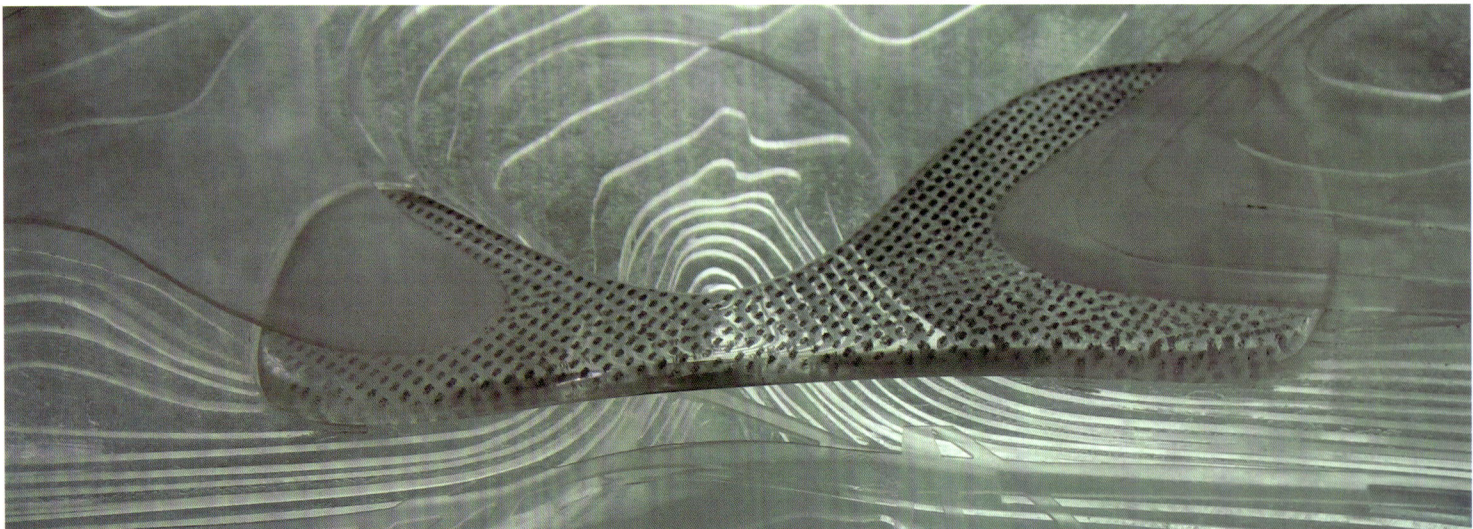

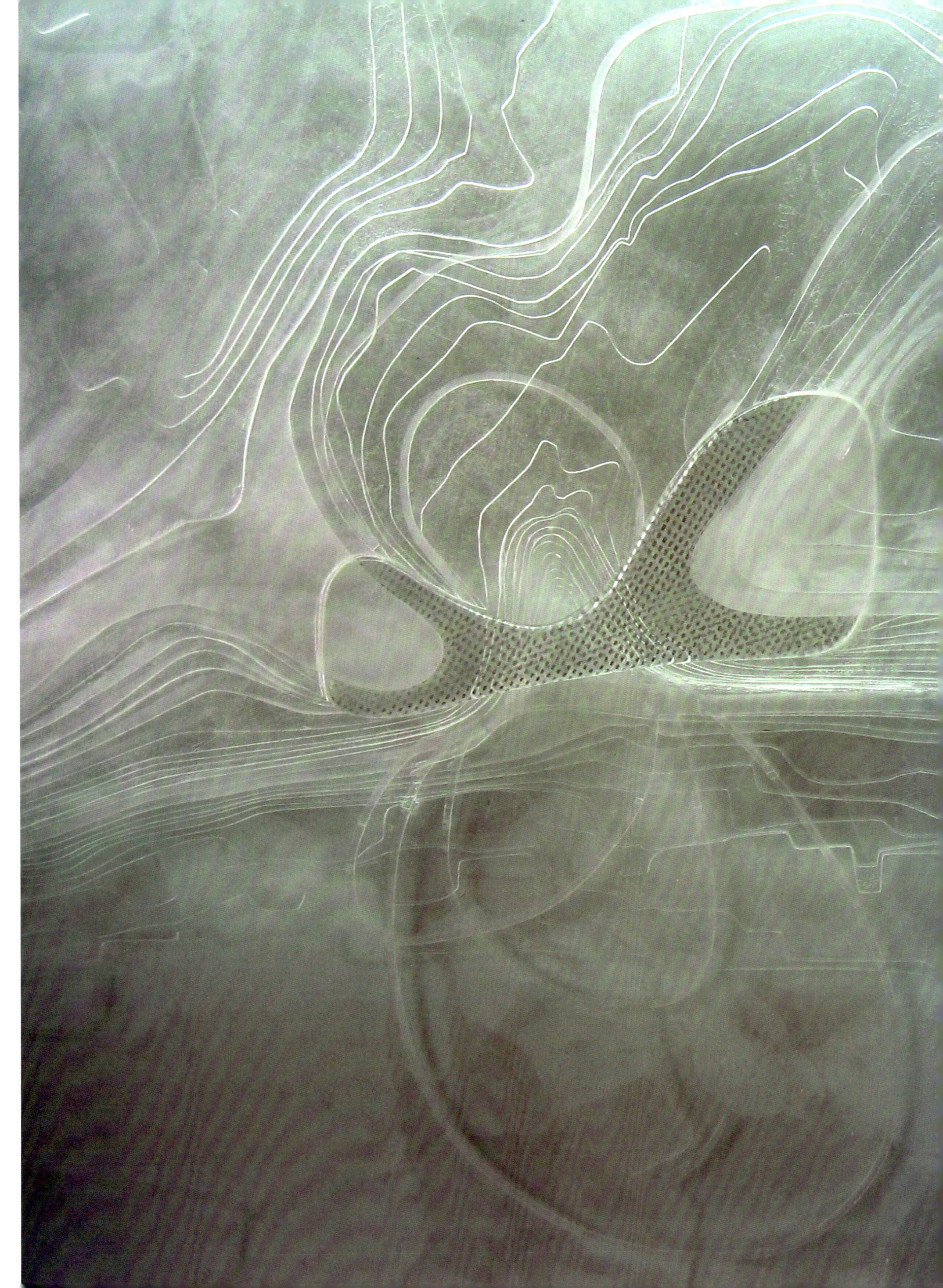

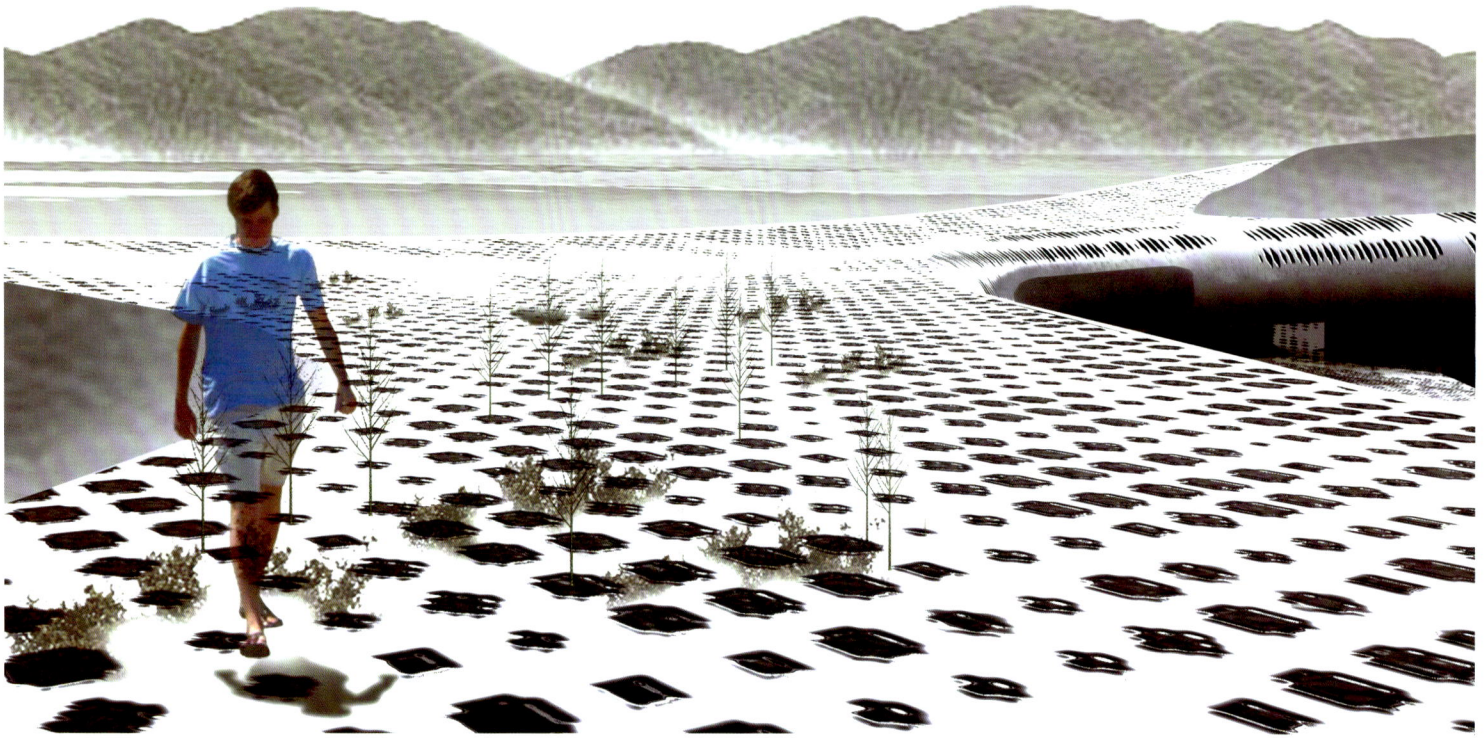

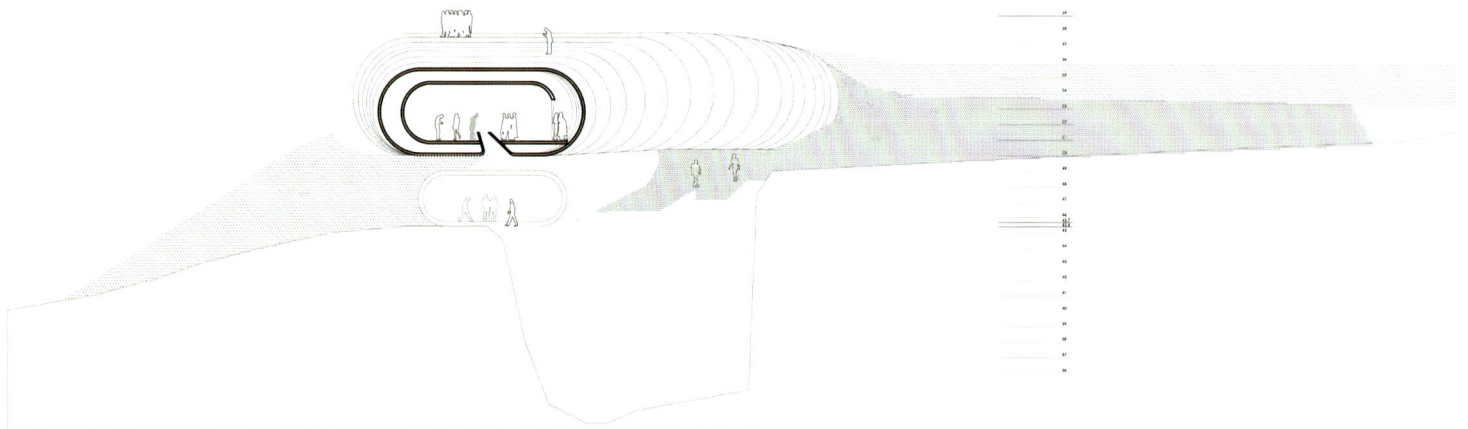

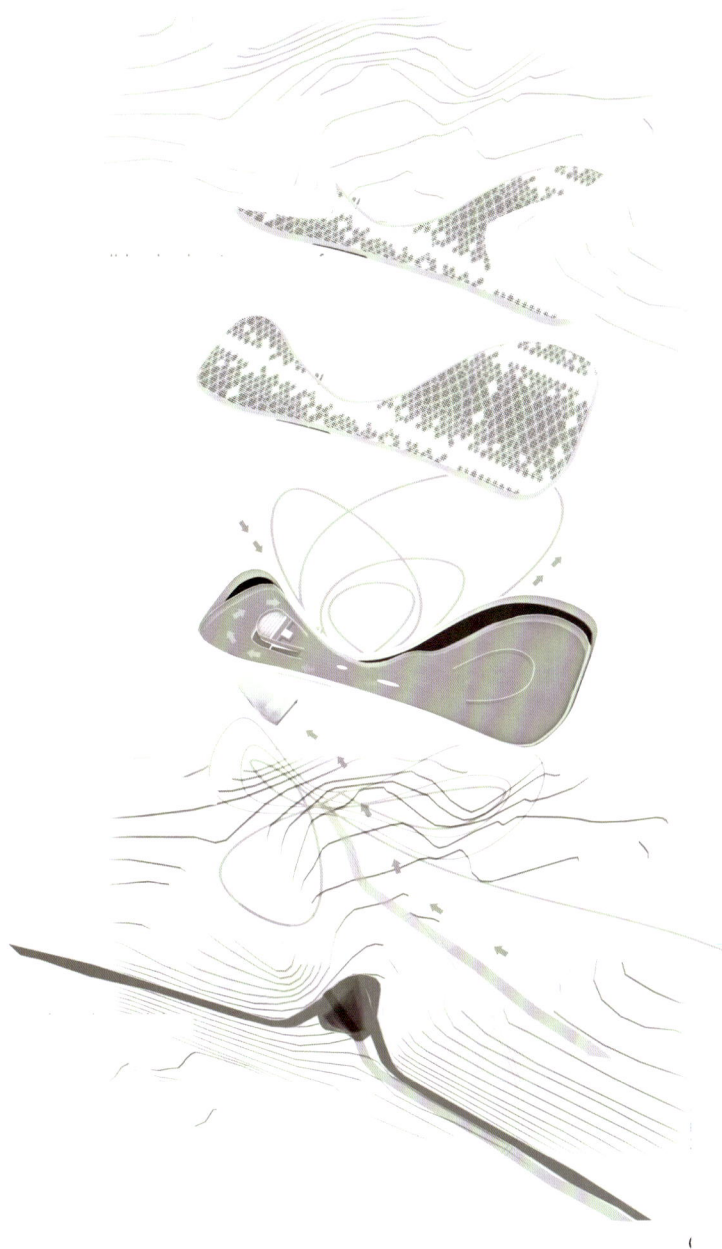

Design concept 设计理念

Aerial view 鸟瞰图

Interior exhibition 室内展区

保罗·普锐斯那 参赛项目2（二等奖）

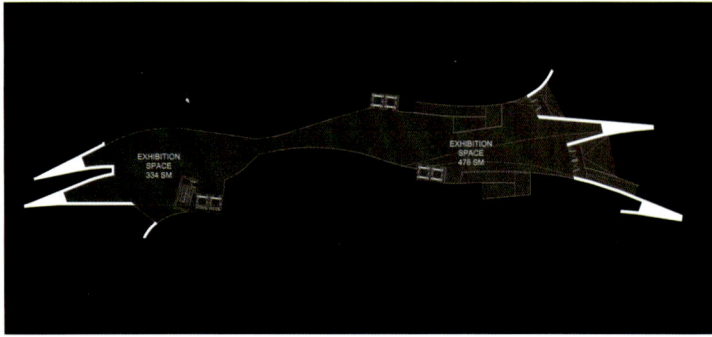

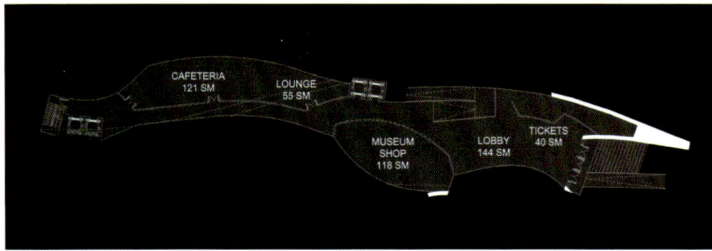

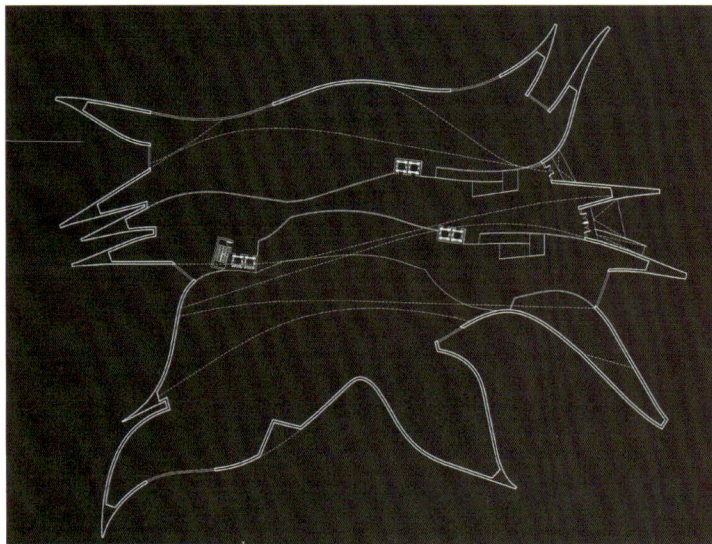

Plan 平面图

Paul Preissner Project 2
the second prize

The design of the second place winner, Paul Preissner consisted of flowing irregular strips containing various programs, a visual mix of previous Peter Eisenman schemes with a splash of Zaha Hadid. In the architects's words: The design proposes an open interconnected terracing of exhibition spaces.

All rooms of exhibition are separated only through the platform circulation, while visibility is constantly maintained. This new form of curatorial framing allows the entire contents of the museum to be constantly appreciated from every position, reinforcing the magnitude of historical artifacts contained within. The site of the building occupies the negotiation between extremes in topography. This allows for the landscape to appear uninterrupted by the project and to appear as though site and building are beautiful confluences of natural forces.

保罗·普锐斯那的设计由漂浮的不规则条状物组成，是艾斯曼和哈迪德设计外观的混合。建筑师称，这一设计让各个阶层的展厅自由的相互连接。所有的展厅之间只是用平台区分开来，造成一种视觉上的连续性。这种新型的展示结构让人从各个角度都可以欣赏博物馆的连续性，并且让馆藏的史前古器物看起来更明显。建筑的场地融合了地质学的极端。这一设计既不破坏自然风景，又将建筑和风景完美的结合在一起。

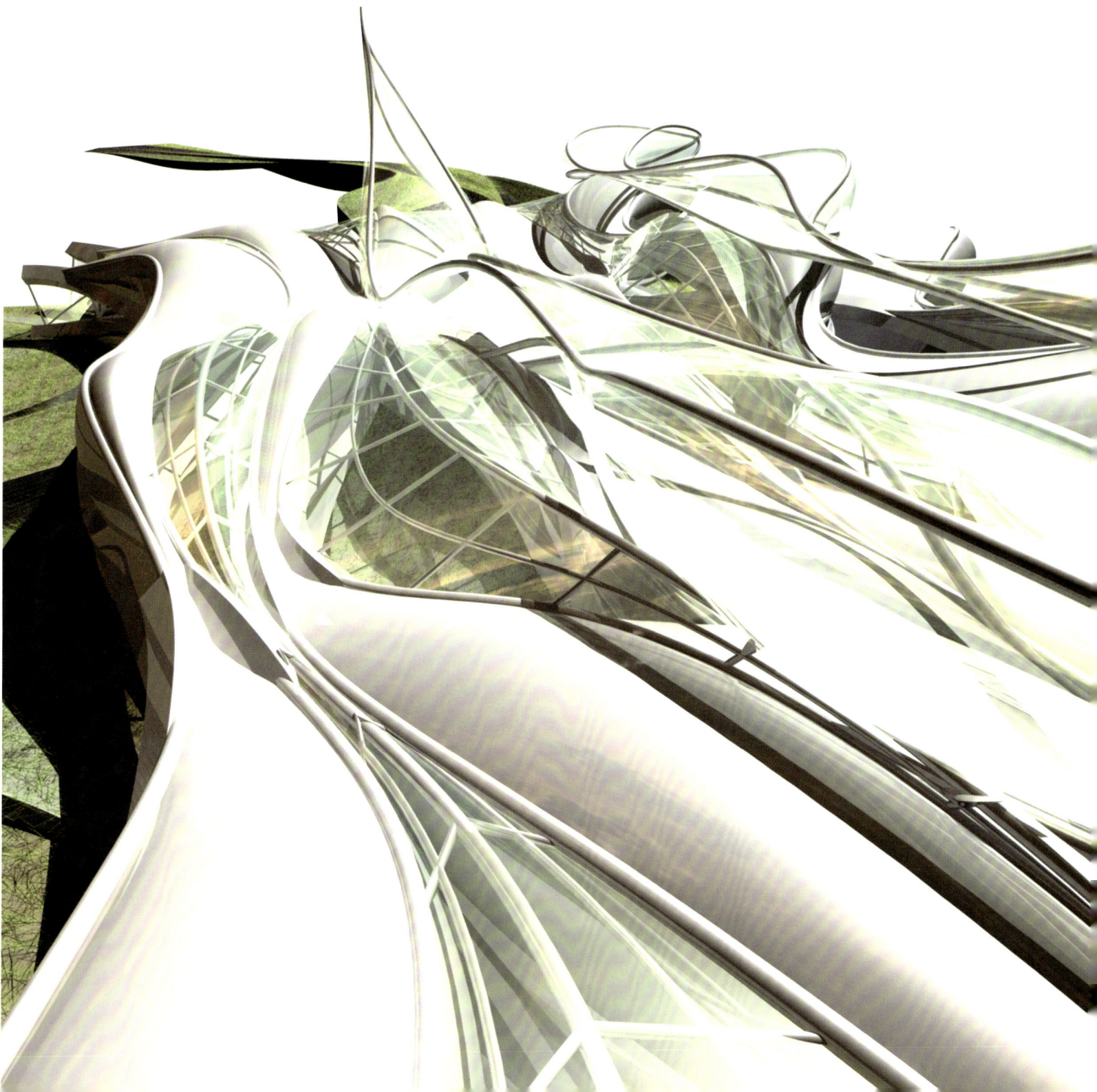

Aerial perspective 鸟瞰图

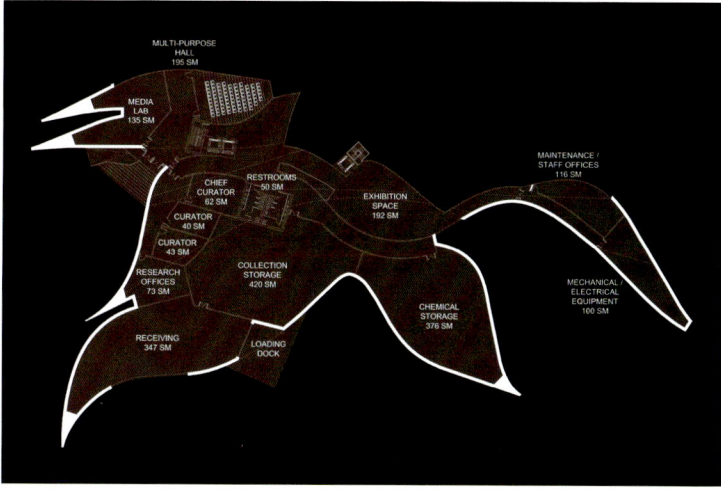

MULTI-PURPOSE
HALL
195 SM

MEDIA
LAB
135 SM

CHIEF
CURATOR
62 SM

RESTROOMS
50 SM

CURATOR
40 SM

EXHIBITION
SPACE
192 SM

MAINTENANCE /
STAFF OFFICES
116 SM

CURATOR
43 SM

RESEARCH
OFFICES
73 SM

COLLECTION
STORAGE
420 SM

MECHANICAL
ELECTRICAL
EQUIPMENT
100 SM

RECEIVING
347 SM

LOADING
DOCK

CHEMICAL
STORAGE
376 SM

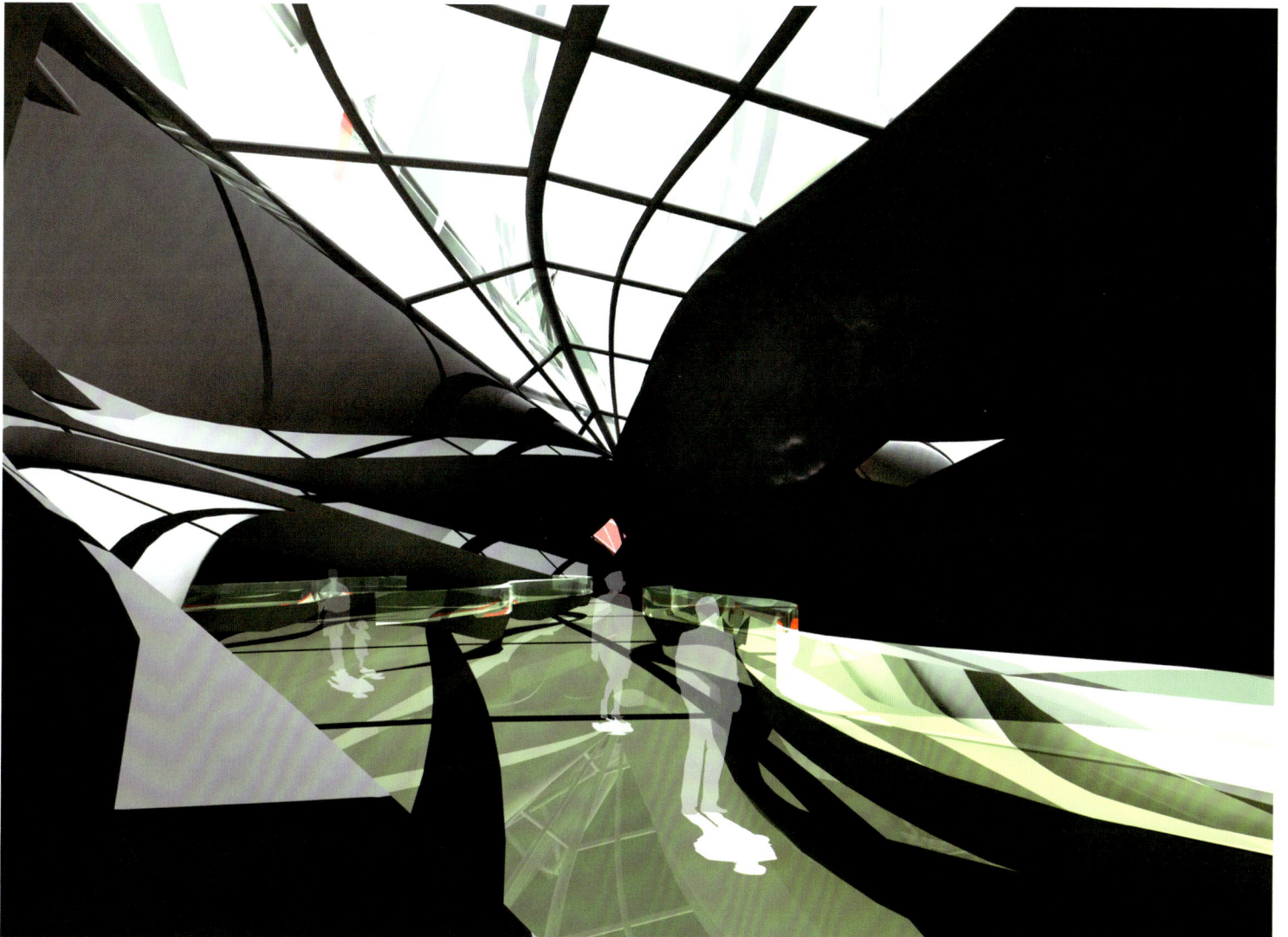

Interior perspective 室内景观

伊斯顿+康博斯 参赛项目3（三等奖）

Easton + Combs Project 3
the third prize

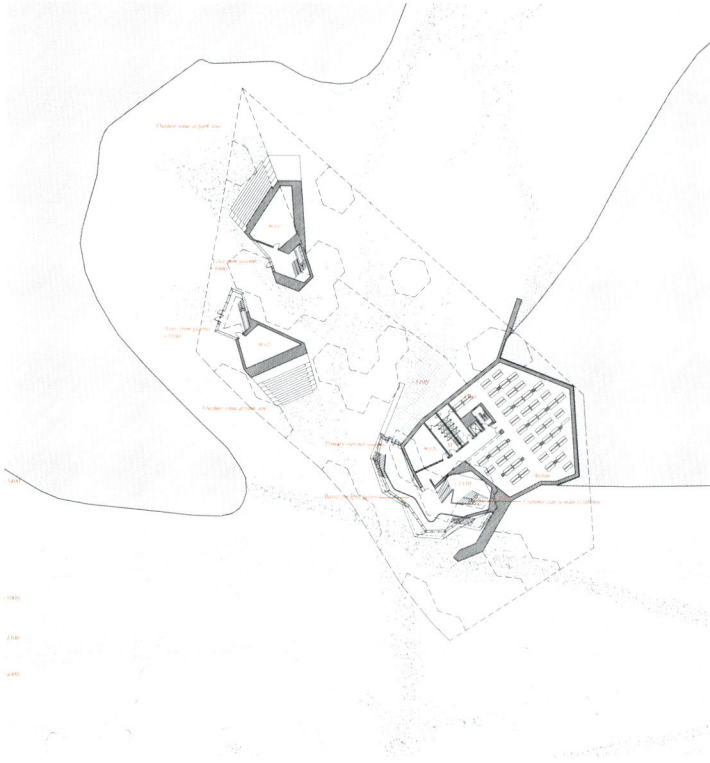

Easton Combs third-place entry also bridged the gap, but in a more compact, conventional manner. The rooftop exterior of their structure somewhat resembled a scatter version of Walter Netsch's "field theory" of geometrical forms, serving as skylights for bringing daylight into the building. This theme is more evident at the next level when the facade is stripped off to reveal the superstructure. The firm's general approach to siting and design was implicit in their statement:

The several hundred artifacts of Achuelean and proto-Acheulean hand axe technology are housed in a series of suspended galleries cut diagonally through the building connecting ground to sky. These galleries reside in a larger constellation of aerated chambers allowing natural light to travel through the main exhibition hall and reach the park surface below. This strategy expresses the building as a suspended inhabitable surface thus reducing the scale of the footprint in the fragile ecology of the site and allowing contiguity of park space between the site of the museum and the pre-history park.

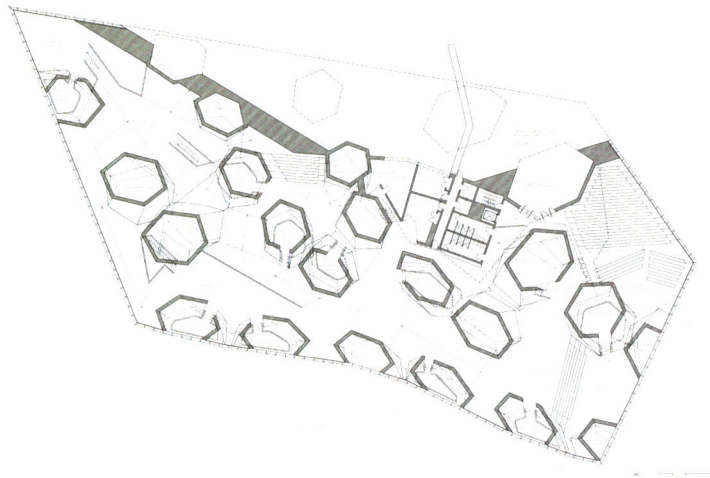

伊斯顿+康博斯的作品同样连接了两座悬崖，但是他采取了一种更为简洁、传统的模式。他设计的屋顶类似于瓦特·内奇的"场论"几何形状，透明的天花板可提供自然采光。剥去建筑的外墙后展示建筑的主体结构，这一设计的主题更为明显。

以下是建筑师对他们设计的阐述：数以百计的阿舍利时期的史前古器物和原始阿舍利手斧技术将在一系列的悬浮展厅内得以展示，这些展厅呈对角线分割，连接天空与大地。天花板上的充气模块让自然光线照射到主展厅和下面的公园。这种设计让博物馆悬浮在地面上，减少了人为对脆弱的生态系统的破坏，维持了公园和博物馆以及史前公园的连续性。

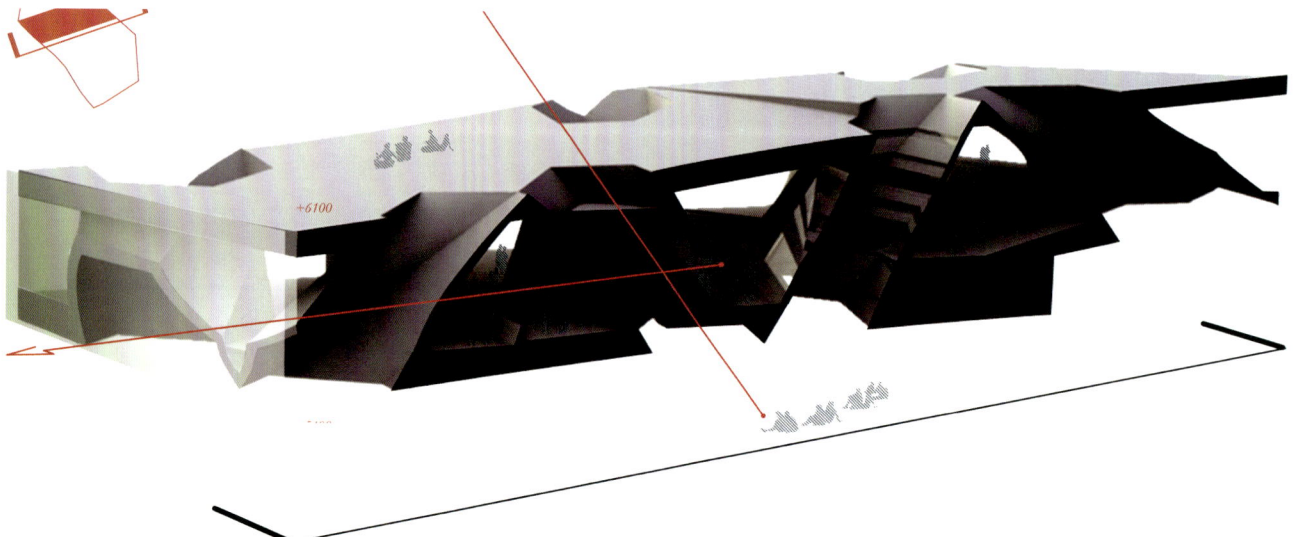

Section 剖面图

Site section 总剖面图

Outdoor Venue — Primary Gallery C — Primary Gallery B — Primary Entrance — Precipice Gallery — Primary Gallery A

Mammoth and Permafrost
Museum Competition
西伯利亚猛犸象和冻土博物馆竞赛

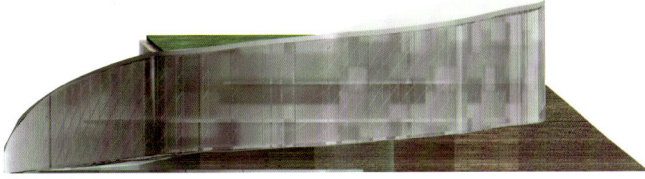

竞赛背景　Competition Background

Yakutsk decided to build a World Mammoth and Permafrost Museum, staging an international competition for the design of the facility. Because of the highly unusual environment/climate conditions, the design challenged faced by the architects in this case were formidable. Contrary to conditions in Alaska, where the permafrost is only a few feet in depth, the permafrost in Siberia reaches several hundred feet in depth.

The program called for a museum not only to display artifacts, but also to promote education and research. The museum is to have a level above ground to for display areas and research facilities and one below ground for a permafrost demonstration exhibit. These levels are to be connected by escalators, which are situated so that visitors can catch a glimpse of the unfolding research activities.

雅库茨克决定建成一个世界猛犸象和冻土的博物馆，举办了一次国际设计竞赛。由于环境、气候条件不同寻常，建筑师所面临的设计挑战是巨大的。与阿拉斯加冻土只有几米深的条件相反，西伯利亚的冻土深达数十米。

这个博物馆不仅是陈列文物之用——最近的一个发现表明猛犸象的历史可以追溯到4万年以前——项目的主要目的是推动教育和研究的发展。这个博物馆的结构将分为两层，地上部分是展示区和科研设施，地下部分是冻土的演示展示区域，各个部分由自动扶梯连接，其分布可使观众一眼瞥见正在进行的研究活动。

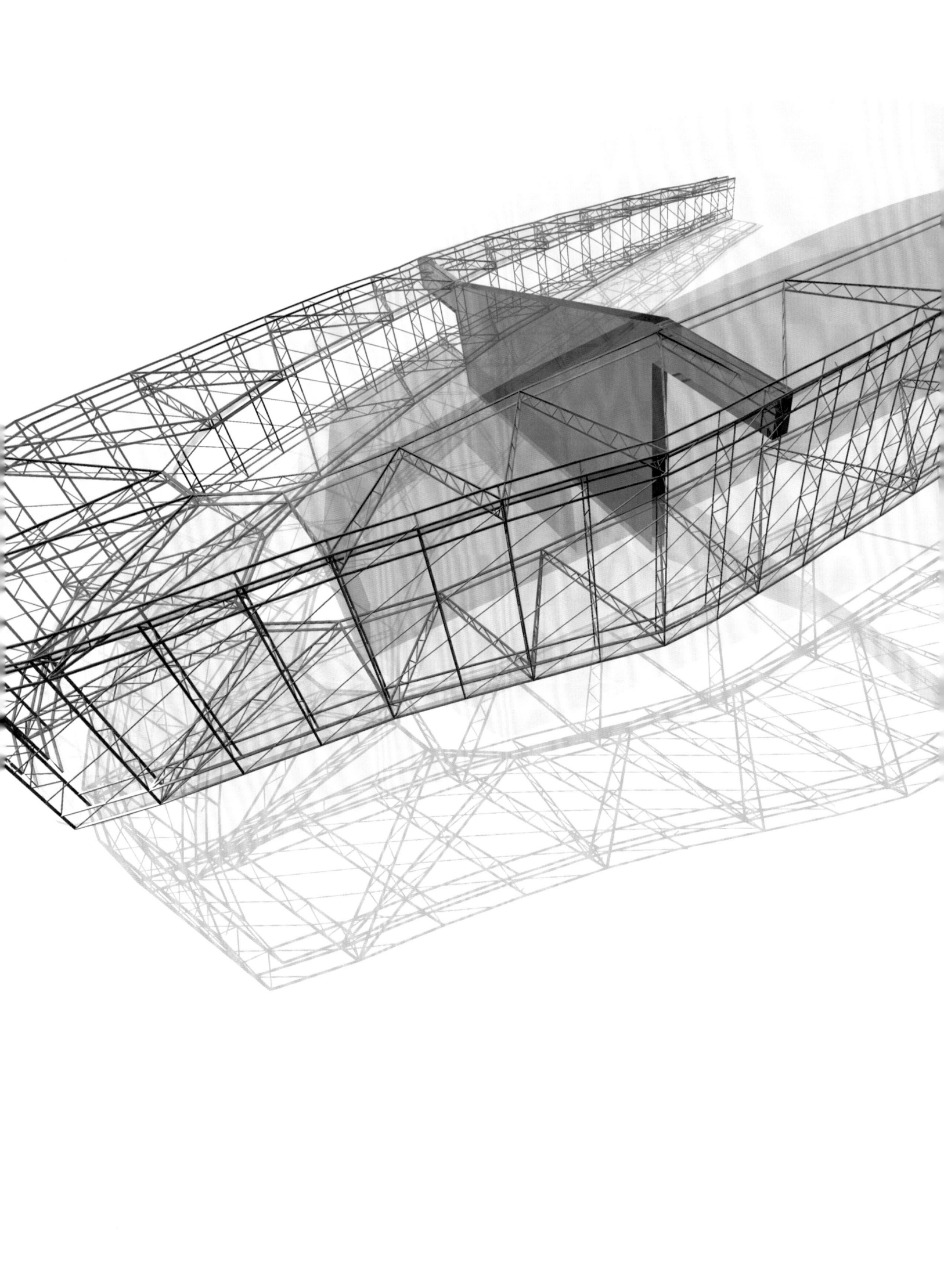

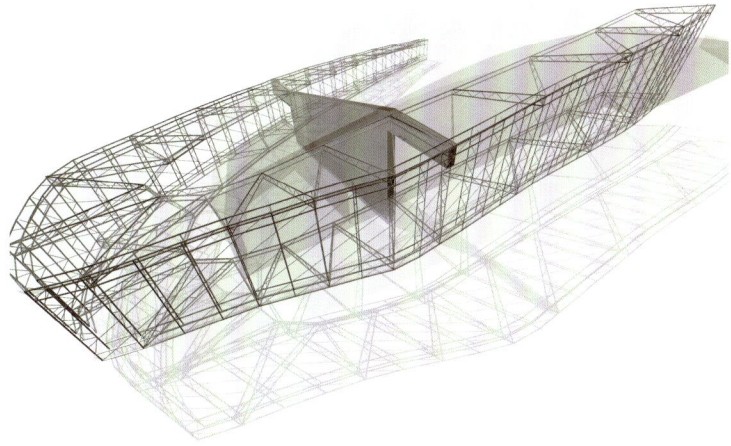

When people go to a museum, they appreciate an experience which can transport them into another world. The design by Leeser would seem to hold that promise.

人们去博物馆时都愿意经历一种进入另一个世界的体验，利舍建筑事务所的设计似乎可以使人们达到这一目的。

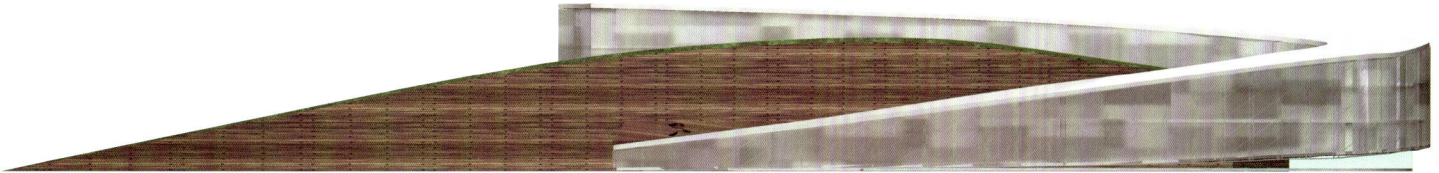

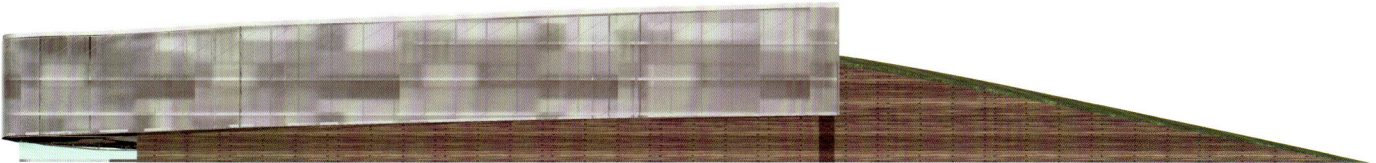

索伦·罗伯特·伦德建筑事务所
参赛项目1（入围奖）

Soren Robert Lund Arkitekter
Project 1 finalist

Though the Danish firm, Robert Lund Arkitekter came up with a design which was notable for its outstanding sustainable qualities, but was less convincing when it came to the creation of a permafrost icon. The scheme's undulating form was more a nod to the water – fine for the summer when it is visible – than creating a sense of mystery surrounding all of those found (and still to be discovered) mammal artifacts from thousands of years ago which carry untold stories with them.

哥本哈根索伦·罗伯特·伦德建筑事务所想出了设计以其出色的可持续性特点为名，但却在创造冻土地标方面缺少信服力。该方案起伏的结构形式更为肯定水的作用——因为夏季可见，所以好看——而不是强调创造一种环境的神秘感，其实，这些几千年前发现的哺乳动物人工制品本身都隐藏着许多不为人知的故事。

Exhibition space 展示空间

Michigan State University Art Museum Competition

密歇根州大学艺术博物馆竞赛

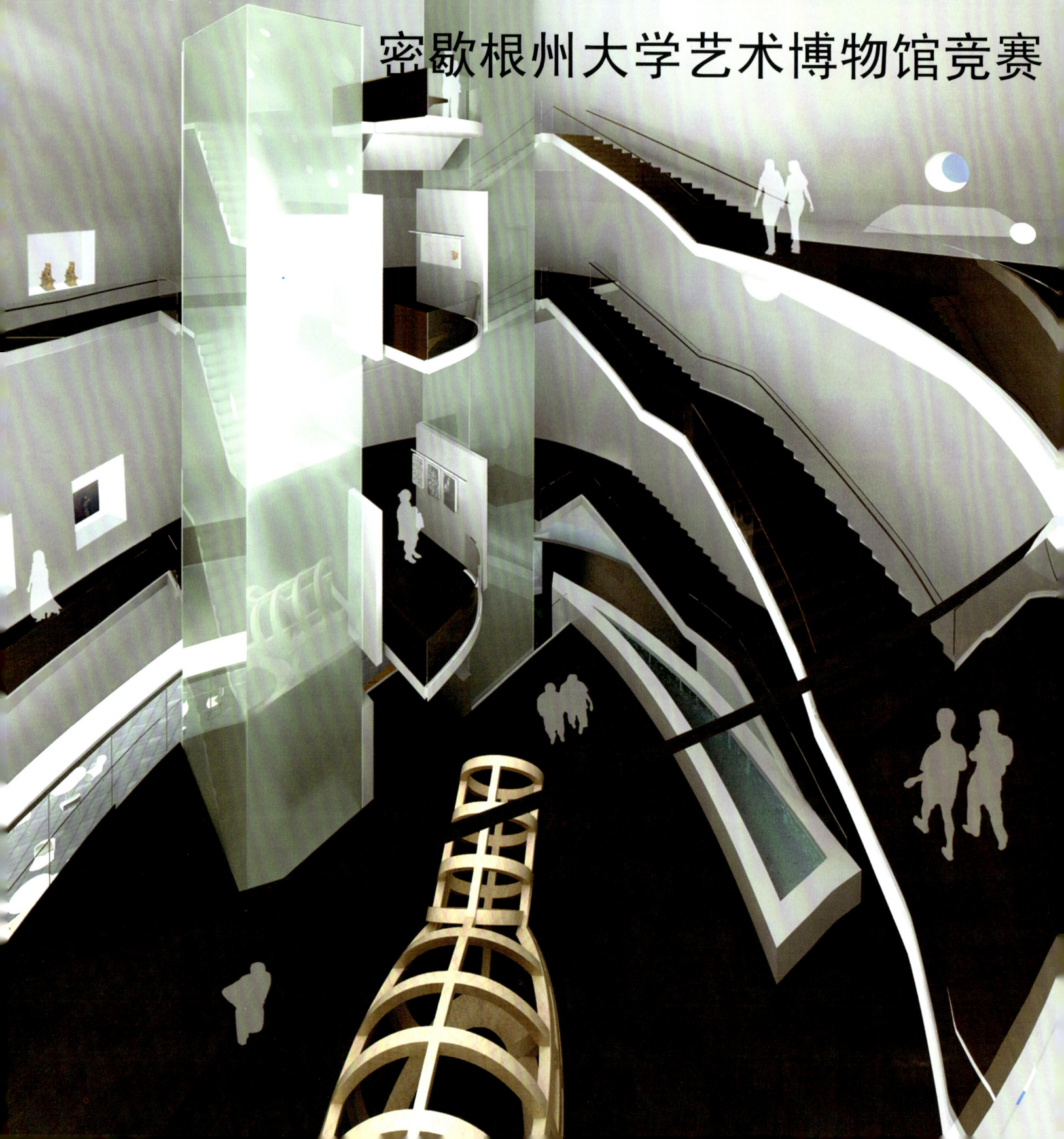

竞赛背景 Competition Background

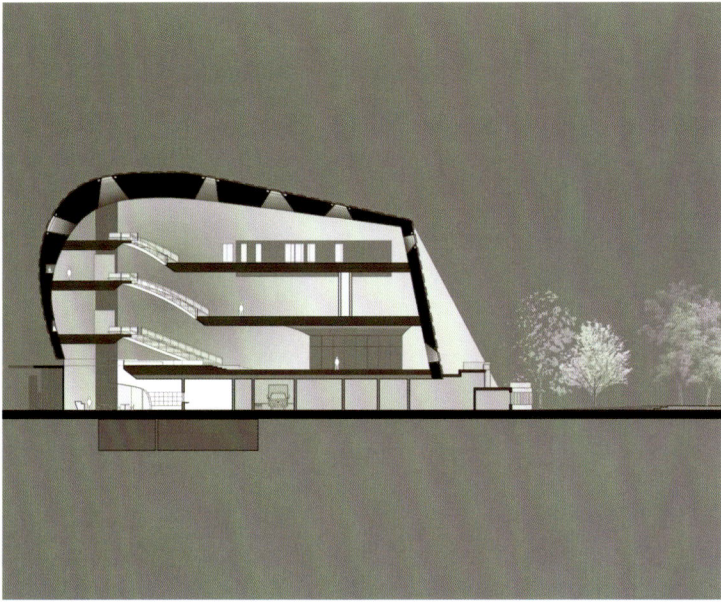

In 2007, Michigan State University was lucky enough to land a $26 million donation for a new art museum from its billionaire alumnus, Eli Broad, and his wife, Edythe. Not content to just pick an architect via the interview process for the museum's design, the East Lansing school then decided to hold a design competition for the structure, which is to be located at the edge of the quiet, leafy campus. The goal was to create an architectural icon that would entice visitors from town and beyond to enjoy the paintings, sculptures and, of course, the building itself.

2007年，密歇根州立大学十分幸运的从富有的校友Eli Broad 和他的妻子Edythe那里获得了两千六百万美元的捐款。密歇根州立大学计划用这笔钱建造一座新的艺术博物馆。不满足于仅仅通过面试的方式来挑选建筑事务所，East Lansing 学校决定举办一次设计竞赛，他们计划把这座艺术馆建在校园安静、树木繁多的角落中。目的是创造一个建筑标志，以此来吸引来自城镇或更远地方的游客到这里欣赏绘画，雕塑，当然，还有这个建筑本身。

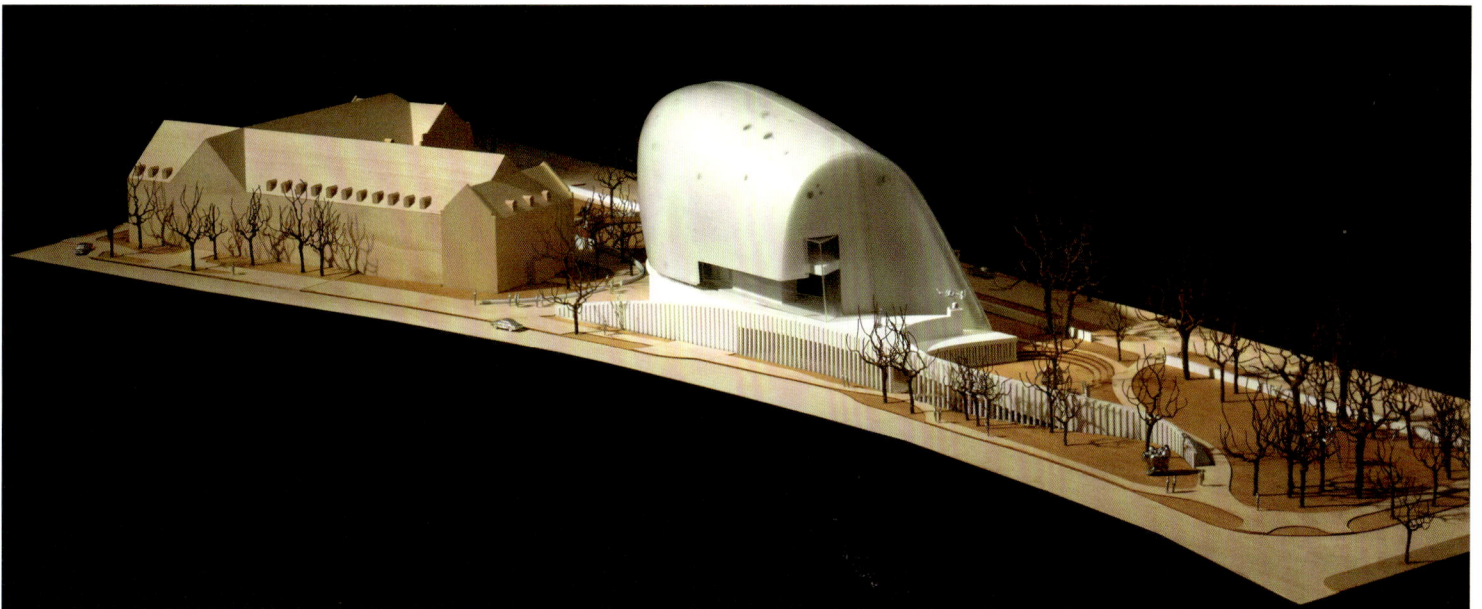

Michigan State has a lot of genteel and graceful buildings but nothing iconic except for a small carillon tower. It is an ideal backdrop for something unique or special. Now it will get that.

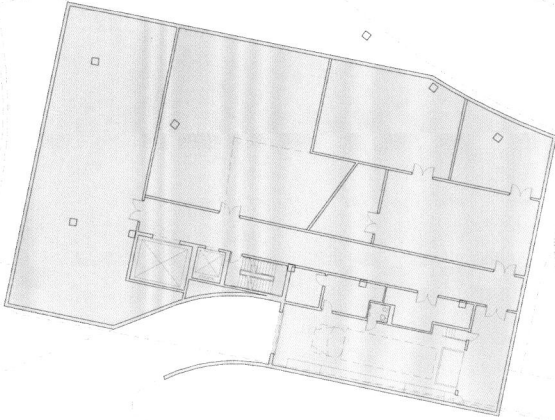

密歇根州有许多时尚和优雅的建筑，但除了一个小的钟琴塔，这里并没有一座标志性建筑。这是建造一个独一无二的建筑的理想背景。现在，这里将会有一个标志性建筑了。

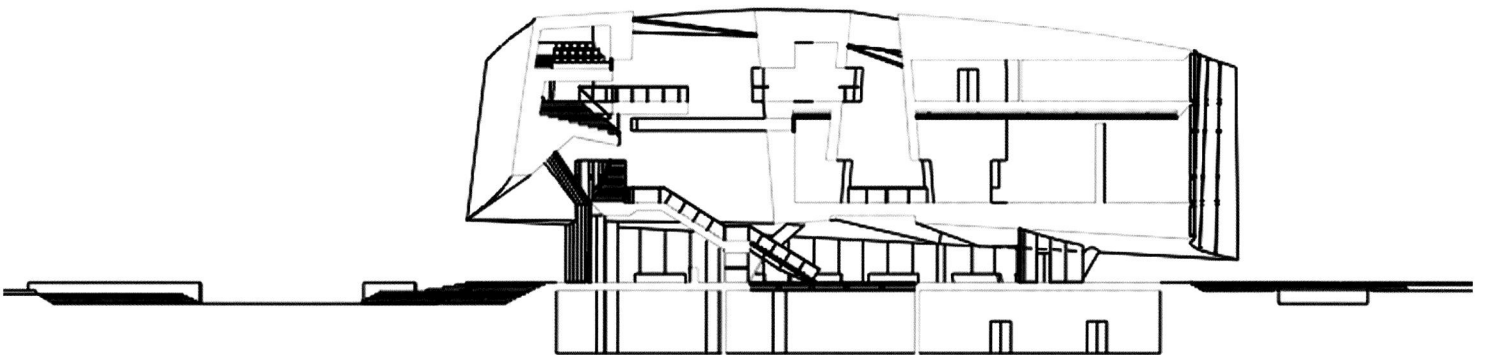

扎哈·哈迪德建筑事务所
参赛项目1（一等奖）

Zaha Hadid Architects Project 1
the first prize

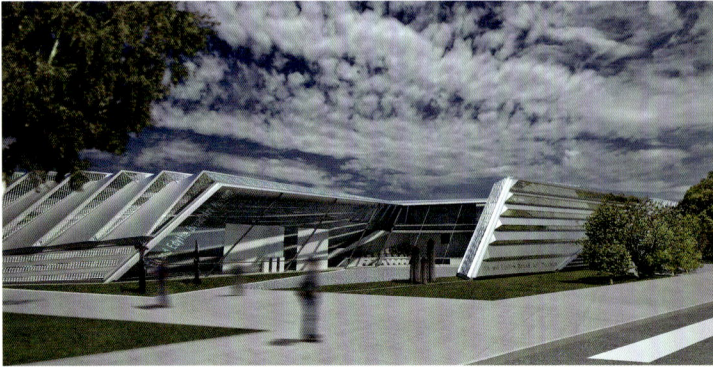

Hadid's design is a long and relatively low-slung trapezoidal structure that tilts 44 degrees at its head and toe walls in a fashion resembling a fast-traveling race car or train. About 315 long and 90 feet wide at its thickest portion, it extends two arms around a courtyard; one extension widens while the other thins to as narrow as 15 feet across. Of its 41,000 square feet total, about 18,000 are for gallery space. Its most unusual and intriguing aspect is its origami-like skin of folded metal and glass in contrasting patterns. The result, Hadid has said, "is a series of colliding spaces." The folds, sometimes transparent and sometimes opaque, offer a peek-a-boo effect to passers-by and museum visitors. That also allows natural light into the museum where appropriate.

Hadid的设计是一个长而低的梯形结构，它从顶部到底部倾斜了44度，这种形态很像一辆高速行驶的跑车或火车。建筑最厚的部分大约有96.012米长，27.432米宽，它伸出的两只手臂环绕着整个院落。一边的延伸变宽的同时，另一边变到只有4.572米宽。在总共3809平方米的面积上，展示空间占据了1672平方米。
整个建筑最与众不同又最有趣的部分是用对立样式的折叠金属和玻璃构成的折纸式外壳。这种设计的结果，用Hadid的话来说，是"一系列的交错空间。"这些折叠部分有时透明有时不透明，为行人和游客提供了一种捉迷藏的效果。这样的设计还使得自然光能进入博物馆中合适的位置。

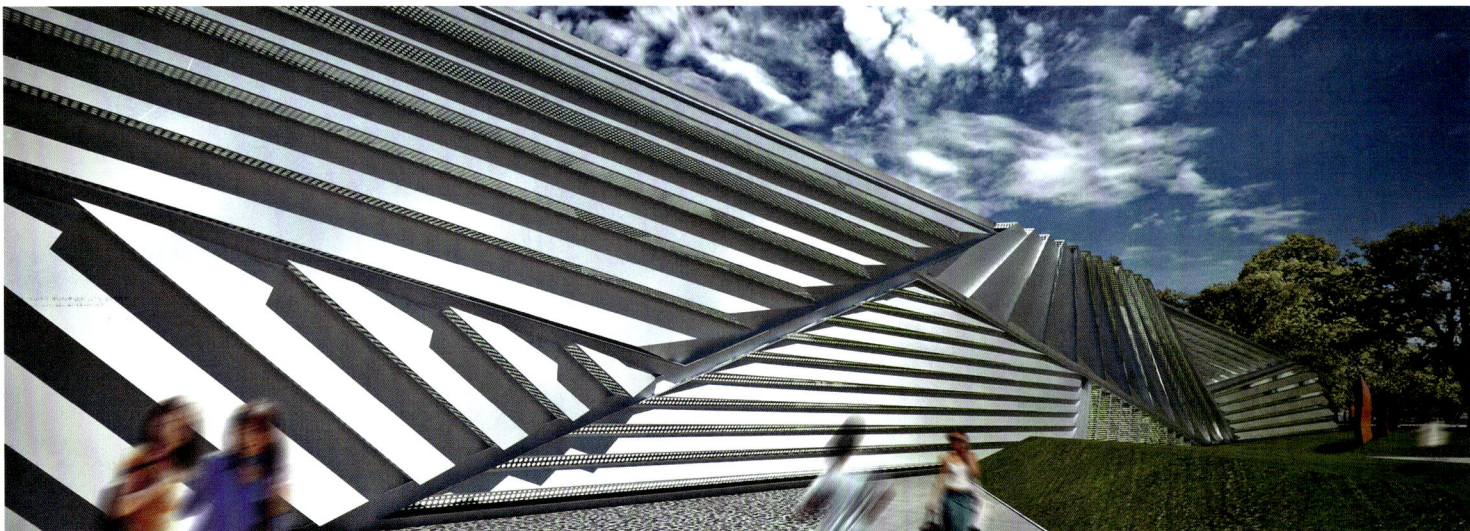

Coop Himmelb(l)au 参赛项目2（入围奖）

Coop Himmelb(l)au Project 2 finalist

Coop Himmelb(l)au's design showed second-story shapes lifting over the lower level and had a dramatic bridge-like walkway and ramp coming down to the sculpture garden.

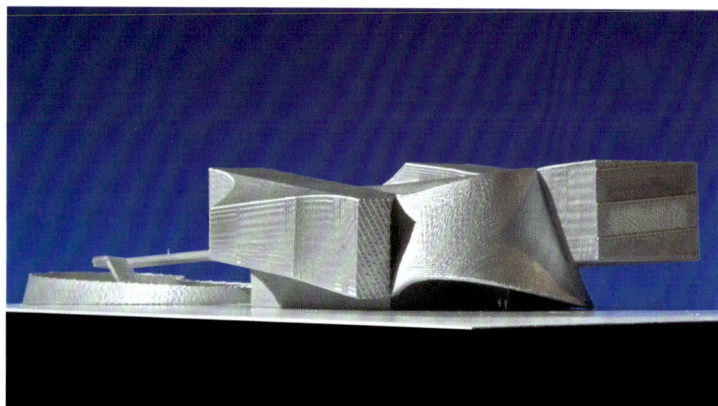

Coop Himmelb（l）au的设计展示了一个二层建筑，它还有一个令人印象深刻的桥型人行道和斜坡，直通向下面的雕塑花园。

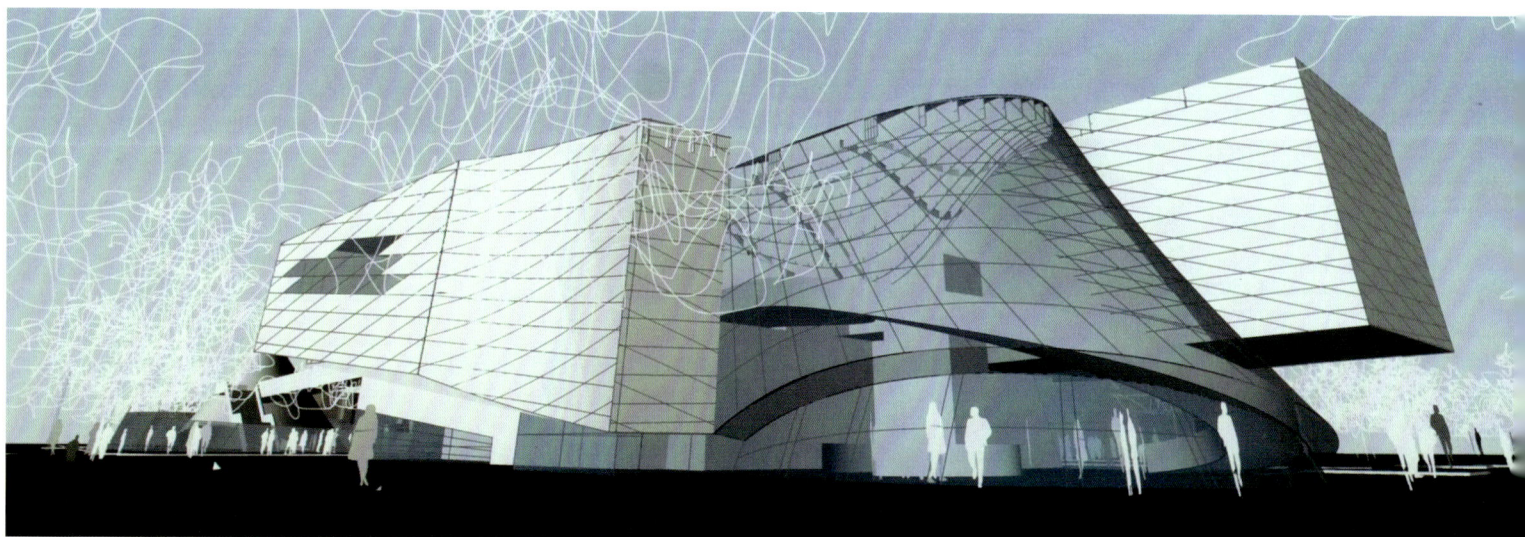

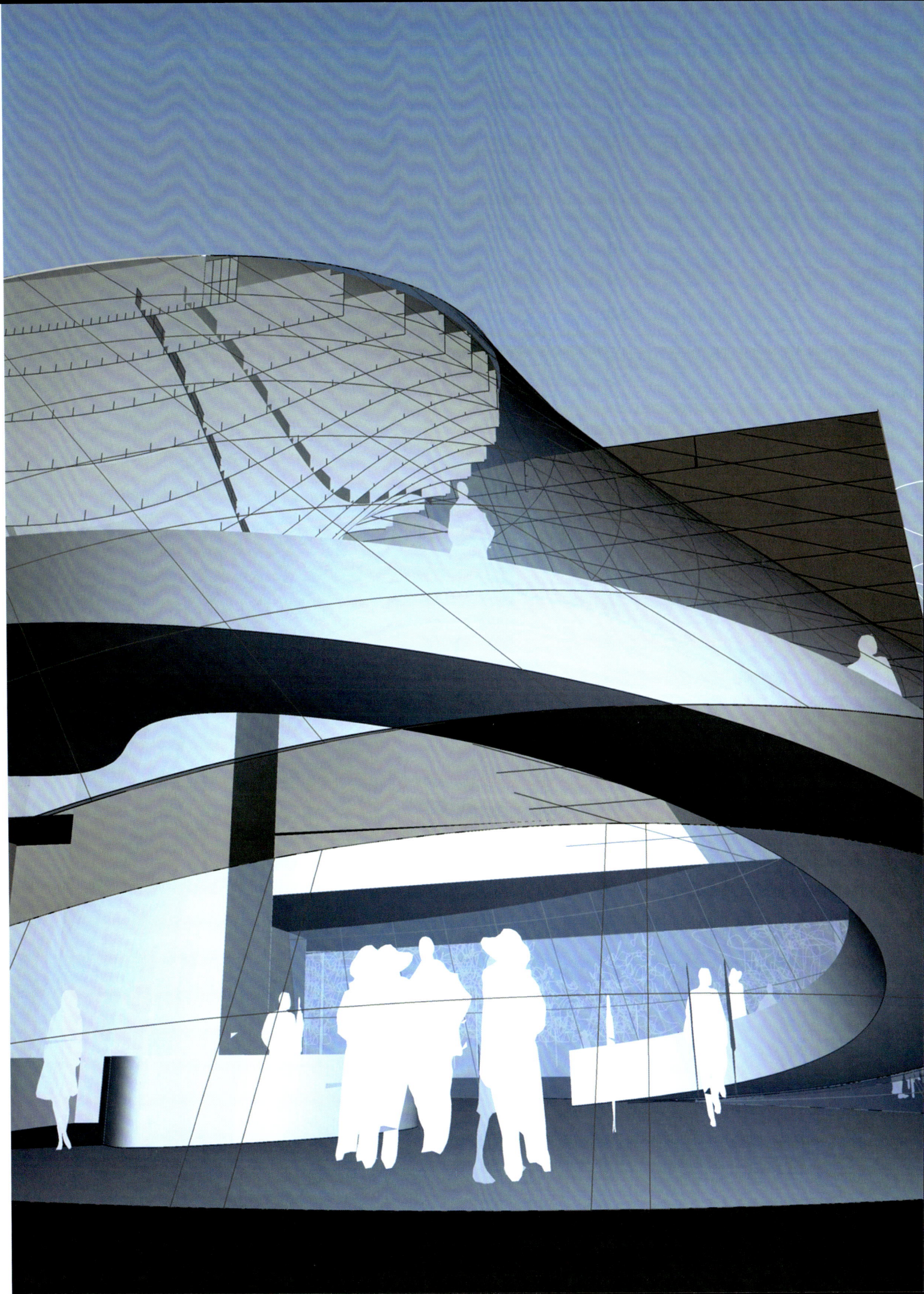

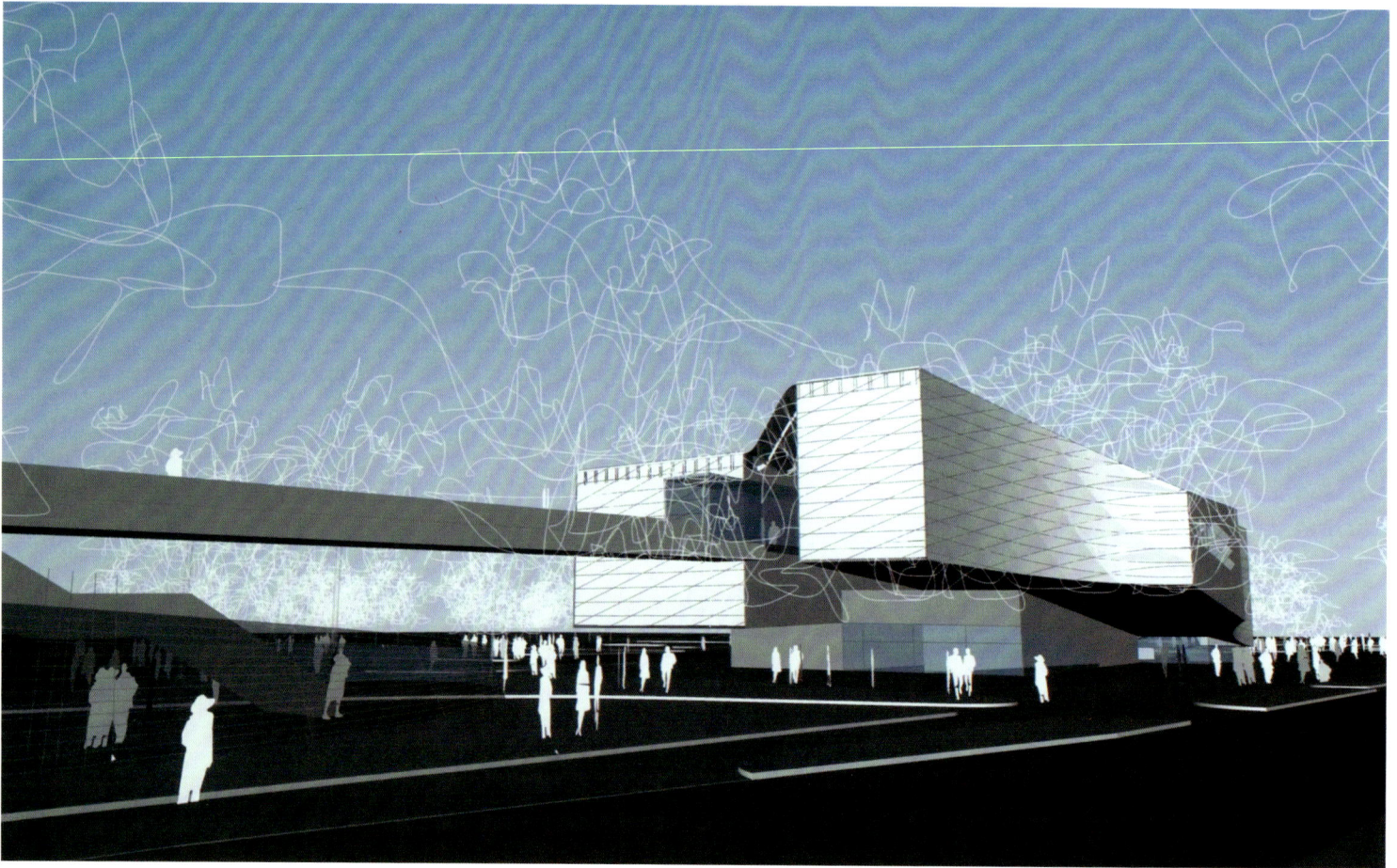

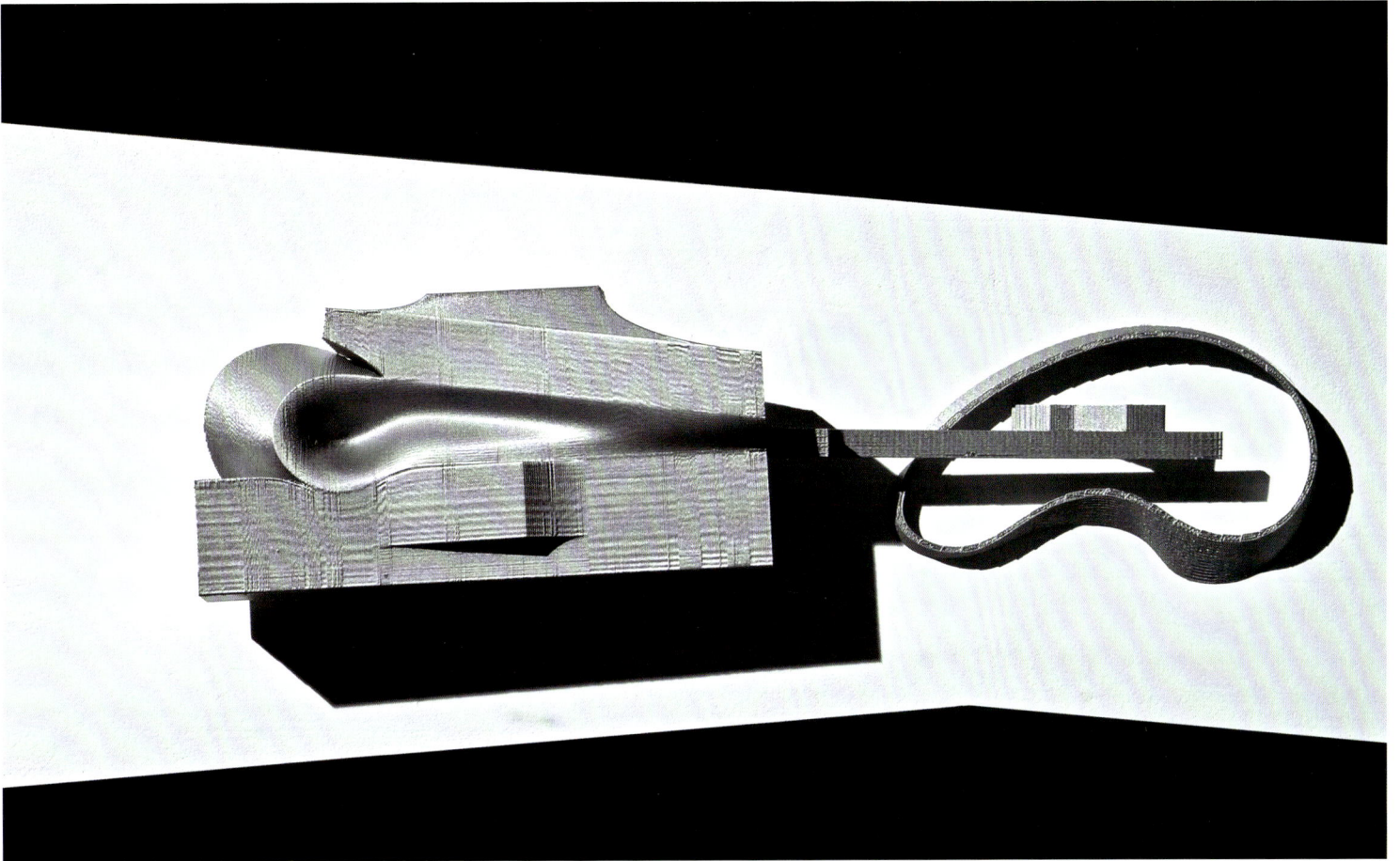

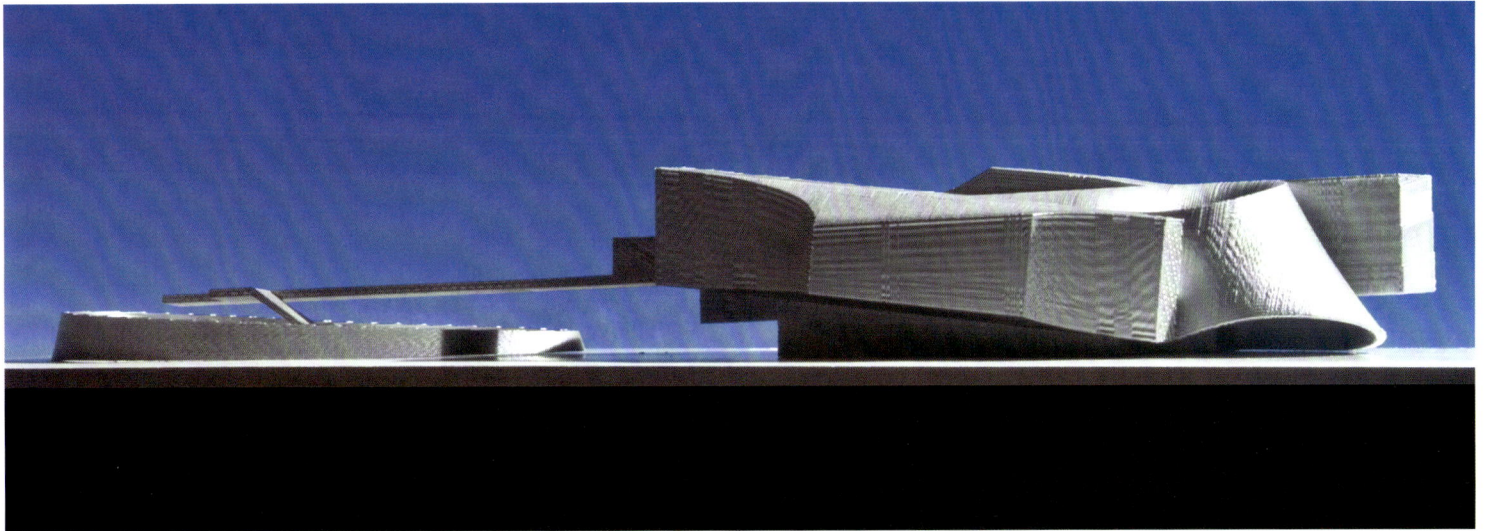

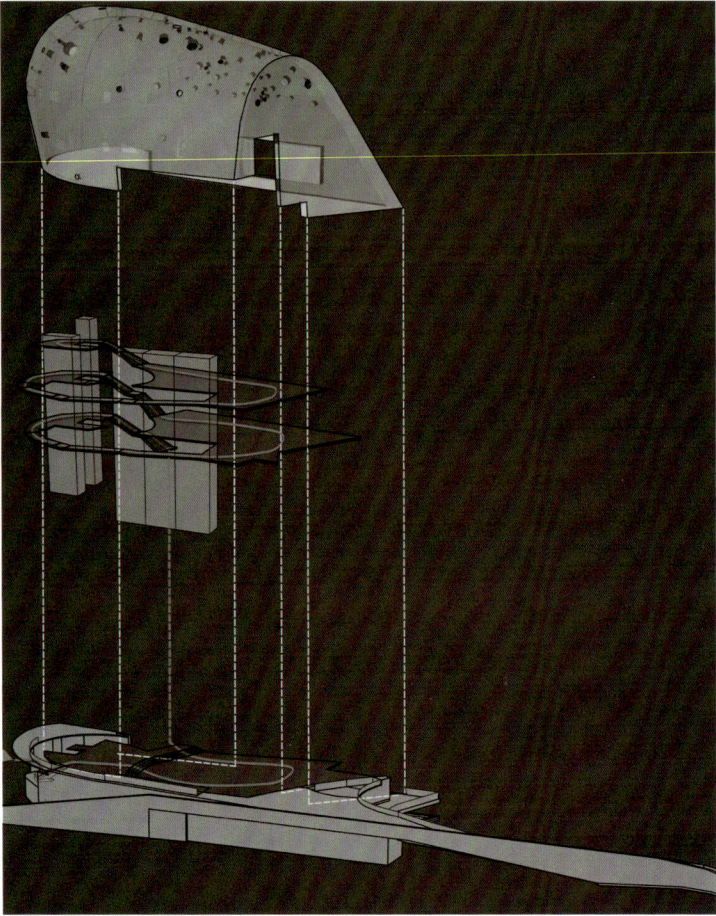

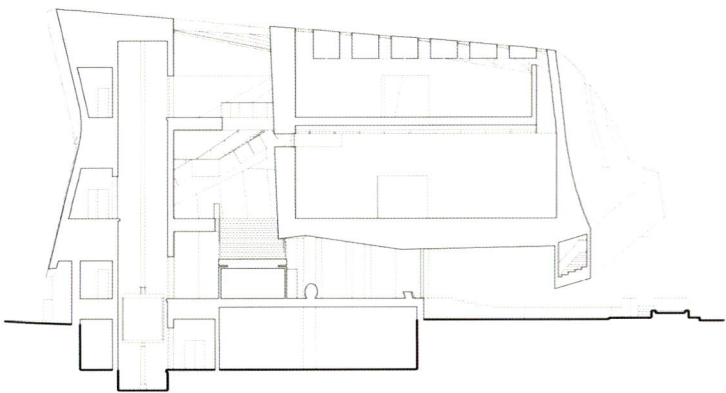

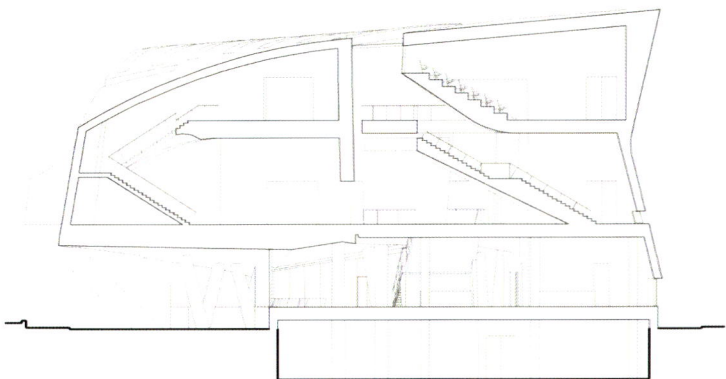

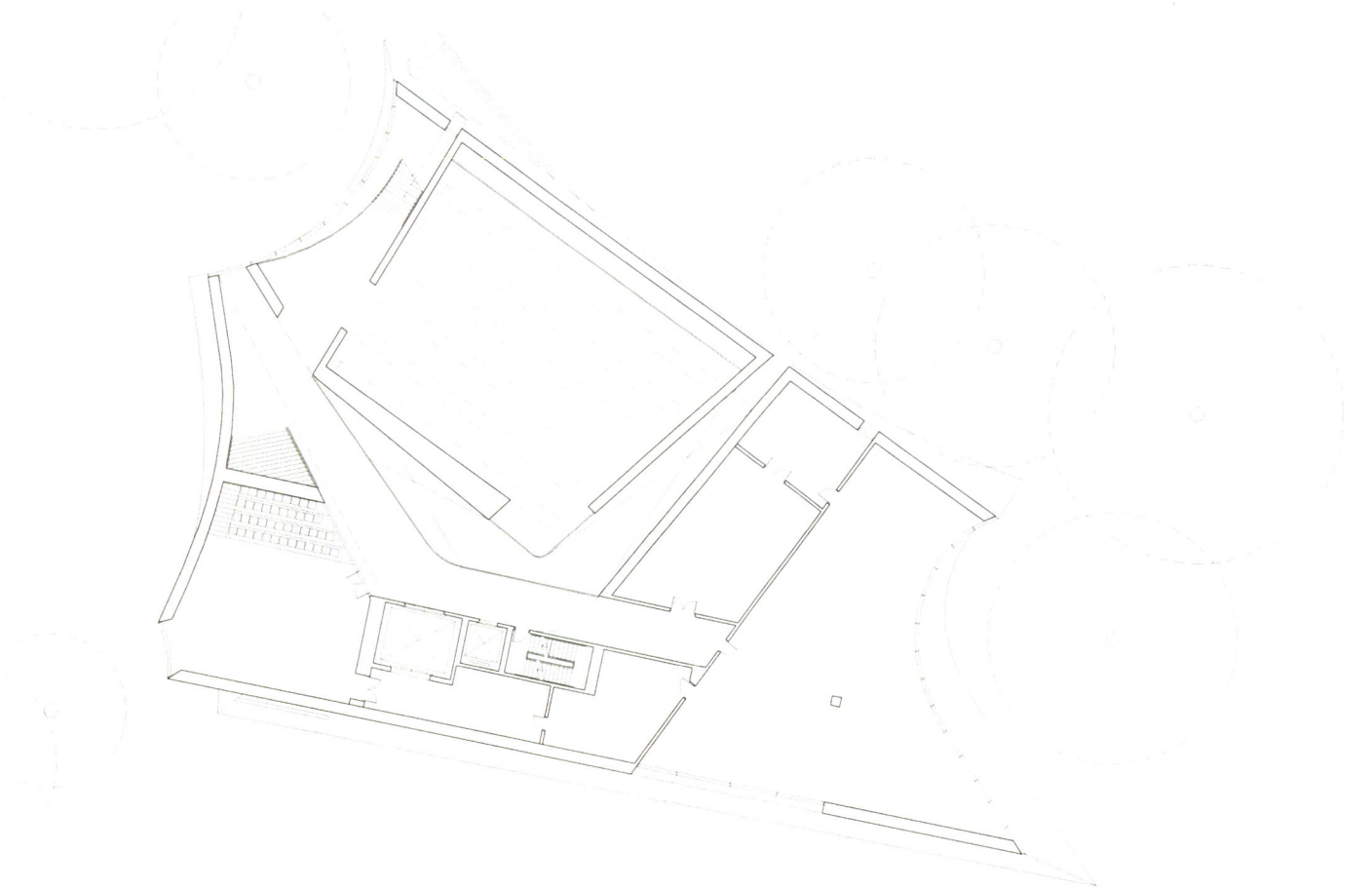

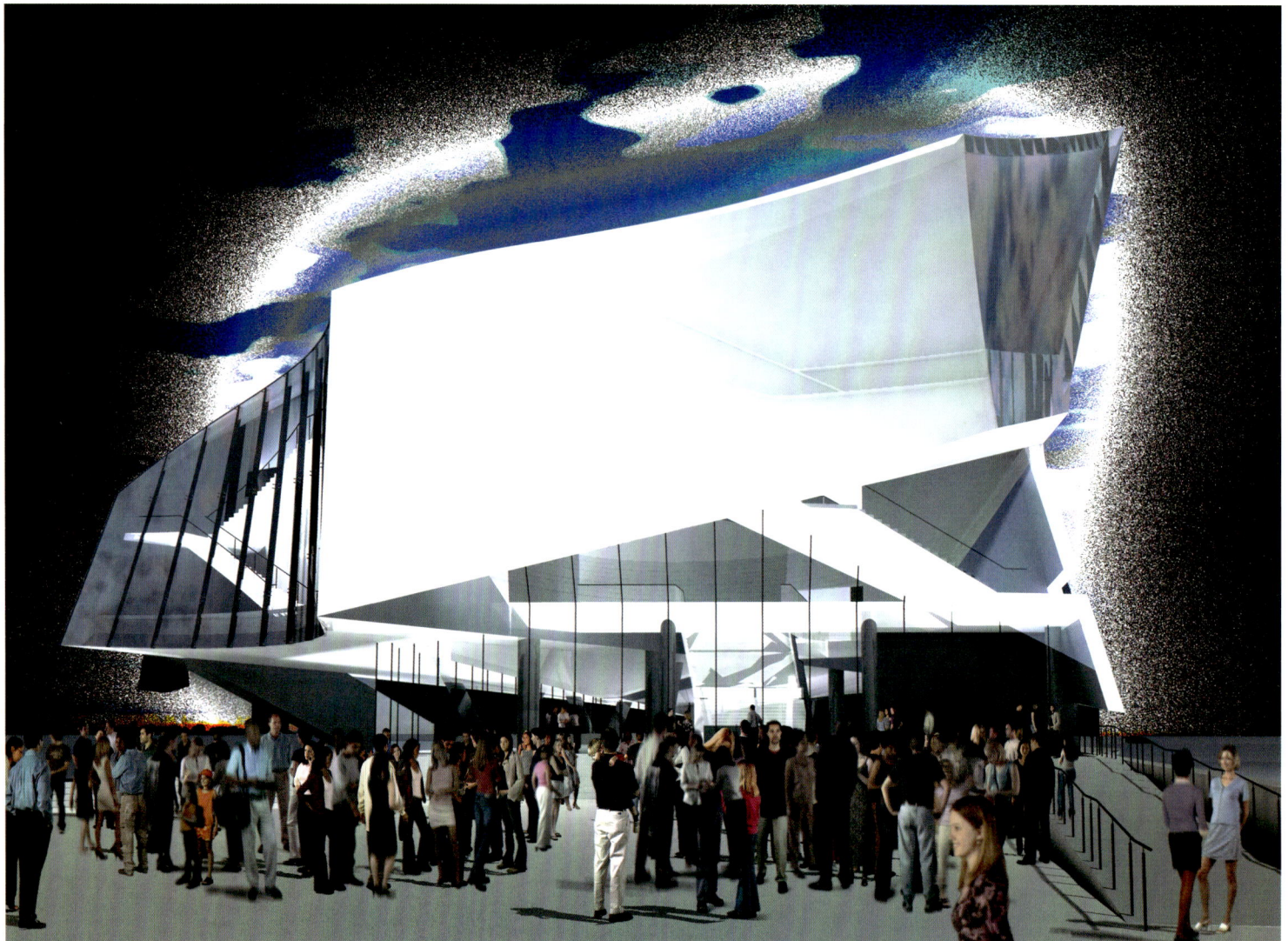

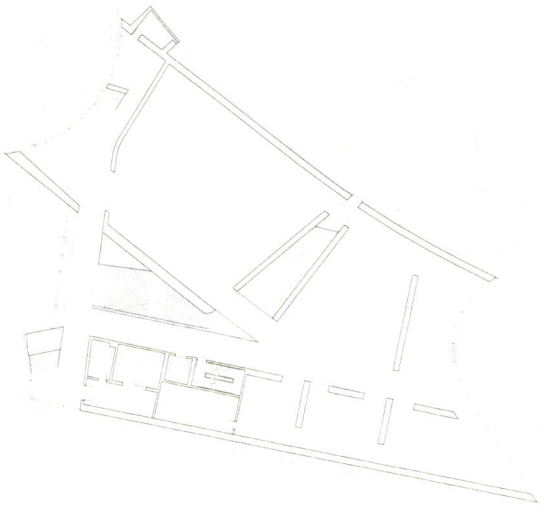

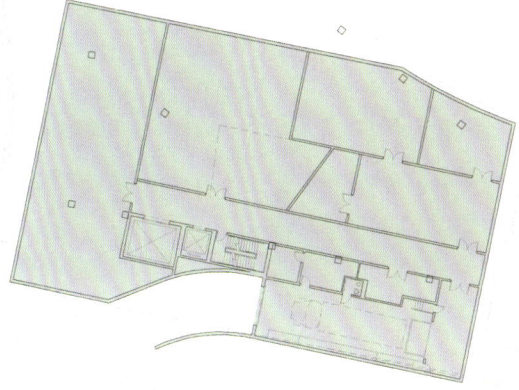

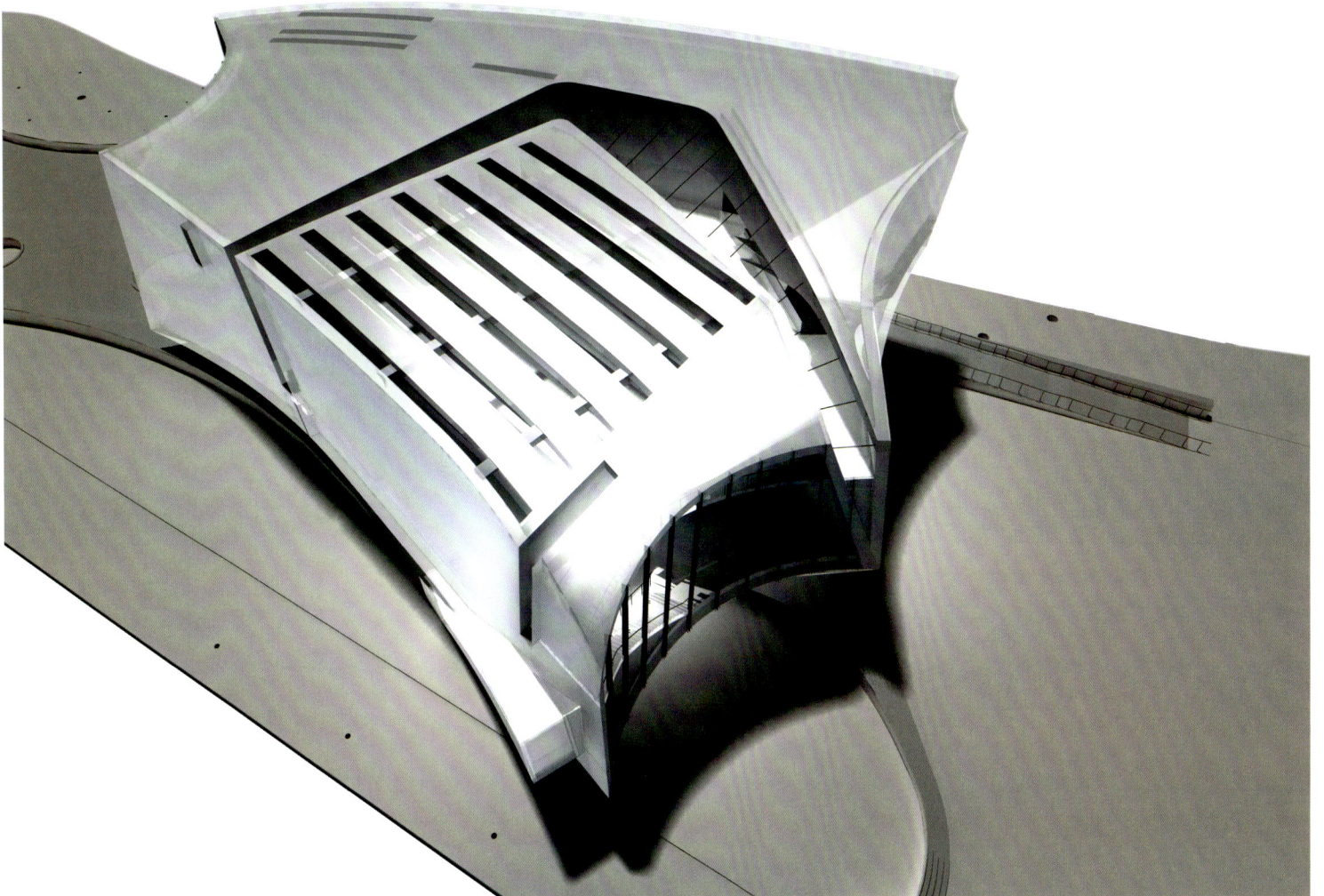

Museum for L'Universitiare catholique de Louvain
Competition
比利时勒芬天主教大学
新博物馆竞赛

竞赛背景 Competition Background

An open, international, two-phase competition was held for the new museum for the Catholic University in Louvain, which is the oldest university in the Low Countries. There is more riding on the new building than just expanded space for the university's local history and art collection. The museum building will be a public symbol of French-speaking Wallonia. The university sought "to offer new access to the city by way of a prominent façade serving as a signal, a future-facing image for the city." Thirty-seven firms answered the call for ideas. The entrants were instructed to address issues of sustainability and to provide two entrances, one from the town and another from the lake.

在勒芬的天主教大学是低地国家最古老的大学，新博物馆不仅为大学历史和艺术收藏所设，新博物馆大楼将是讲法语的瓦隆尼亚的一个公共象征。学校努力通过一个标识性的未来建筑提供城市新的入口。有37家建筑事务所参加了竞赛，参赛者要提出解决可持续性问题的方案，设计2个入口，一个指向城内，另一个指向湖区。

2021st - Paintings 1 (E.2.3.2)
125 sm
(150)

2021st - Paintings 2 (E.2.3.5)
60 sm
(76)

2021st - Paintings 3 (E.2.3.4)
60 sm
(75)

2021st - Paintings 4 (E.2.3.3)
60 sm
(75)

"Eddy Merckx" Room (E.2.3.1)
70 sm
(65)

Terrace

Open to Below

专家点评 Remarks

All in all, the winner's design goes beyond the currently fashionable, for the entire composition is exciting beyond being just a pretty face. Perkins and Will's thoughtful planning will integrate museum, university, and town, making it a true civic center. The power of the proposal resides in the development of a few main ideas, such as the luminosity of the spaces, the perception of the volume and the internal space, the renewable energies and the accessibility of the museum. The competition process has served one small Belgian city very well, and it may have produced one of the better museums of this great museum building period.

总之，获奖方案远远超出现在的博物馆流行水平，因为整个建筑不仅外观好看，而且令人激动不已。帕金斯和威尔建筑事务所/埃米尔·沃哈根建筑事务所的方案的优势在于几个主要观点的发展，比如说，空间的亮度、体积和内部空间的感知、可再生能源和博物馆的可亲性等等。他们的方案将把博物馆、大学和城市整合起来，使其成为一个真正的市中心。

这次竞赛程序成功地满足了一个比利时小城的要求，可能产生出这一博物馆建筑伟大时期的一个更好的博物馆。

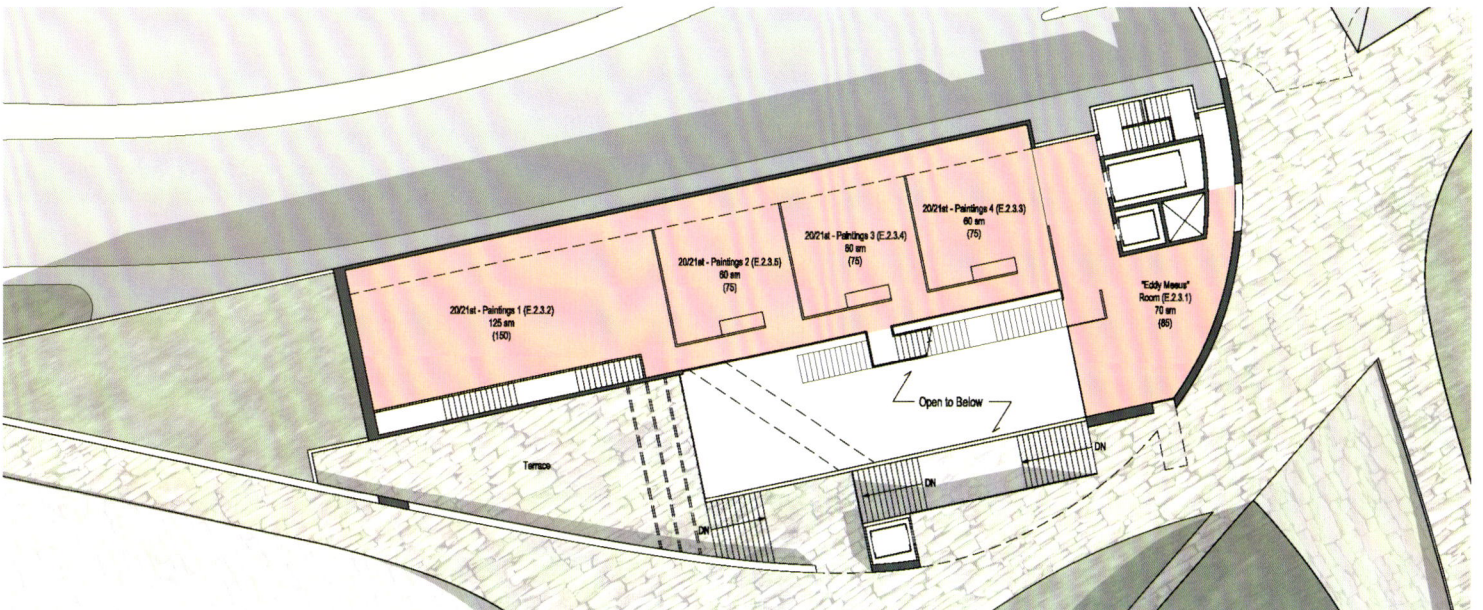

帕金斯和威尔建筑事务所/埃米尔·沃哈根建
筑事务所 参赛项目1（一等奖）

Perkins + Will/ Emile Verhaegen Project 1 the first prize

For this project, Perkins + Will teamed up with Emile Verhaegen of Brussels. The successful scheme covers the entire site, using the building as the city park, with grass roofs that visitors can walk over.

The intriguing Perkins + Will/Verhaegen museum should be an instant landmark. It is composed of two major elements, an exhibition tower and the lower mass of the park covered wing. More than what the architects called a "signal", the four-story tower contains the permanent art collection, office, and workshops. The undulating lower wing houses the more public functions of an auditorium, bar, and temporary exhibition space. Both wings are connected by a large atrium that acts as the museum's backbone, connecting every space and providing continual orientation to the lake. Skylights in the tower create a lively, inviting, and constantly changing space.

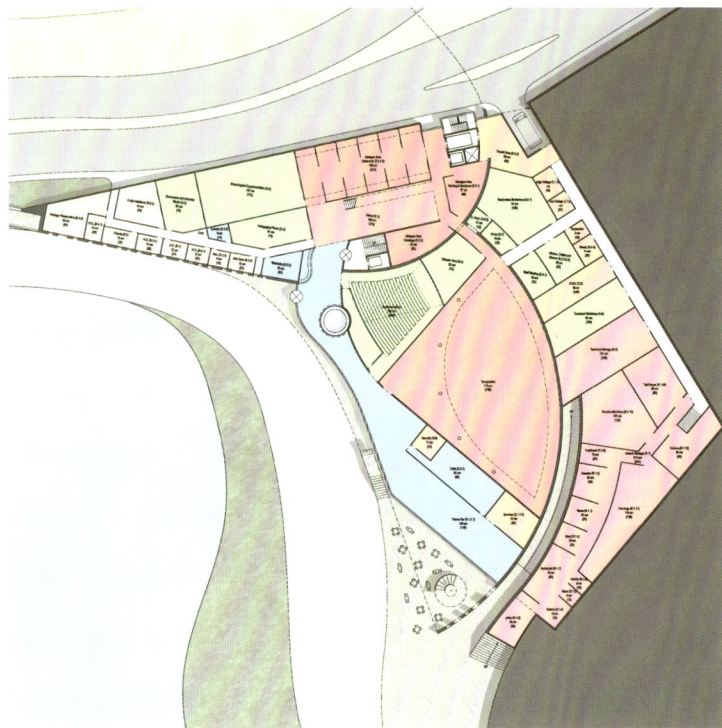

Site plan 总设计图

为了这个项目，帕金斯和威尔建筑事务所与布鲁塞尔的埃米尔·沃哈根建筑事务所联合，这个成功的方案覆盖了整个项目地点，把建筑当城市公园用，上面有参观者可以行走的种有绿草的屋顶。

帕金斯和威尔建筑事务所/埃米尔·沃哈根建筑事务所设计的这个有趣的博物馆应该马上就会成为一种地标性建筑。方案包括两大设计元素，一个是展览馆大楼，一个是有公园遮盖的侧楼。这座4层大楼不仅仅是建筑师口中的"标识"，大楼内部有永久艺术收藏室、办公室、商店和工作室。起伏动感的低层侧楼内设大礼堂、酒吧和临时展览空间，具有更多的公共职能。两个侧楼由一个大型的中庭连接，中庭是整个博物馆的脊梁，连接每一个空间并且向人们提供到达湖区的导向。楼内天窗创造出一个活泼、诱人，不断变化的空间。

View from city 外观图

Aerial view 鸟瞰图

Interior from entrance 从入口处看室内

Interior view 室内

福克萨斯建筑事务所 参赛项目2（入围奖）

Fuksas Architects Project 2 finalist

The Fuksas scheme creates an "artificial landscape" by covering the entire site in a single organic form. Or, in their words, the "surroundings and the over-imposed synthetic landscape interweave and interact with each other; the consequences of this topographic manipulation synthesize the two systems into a consistent progressive and innovative identity." The marshmallow-colored amorphousness of the design looks a lot like Zaha Hadid's Caligari museum project, with cartoonish elements of German Expressionism added in.

In a way, this is the perfect finalist design: it allows the jury to appear really avant-garde while avoiding the controversy and construction headaches that realizing such a fantastic design will entail.

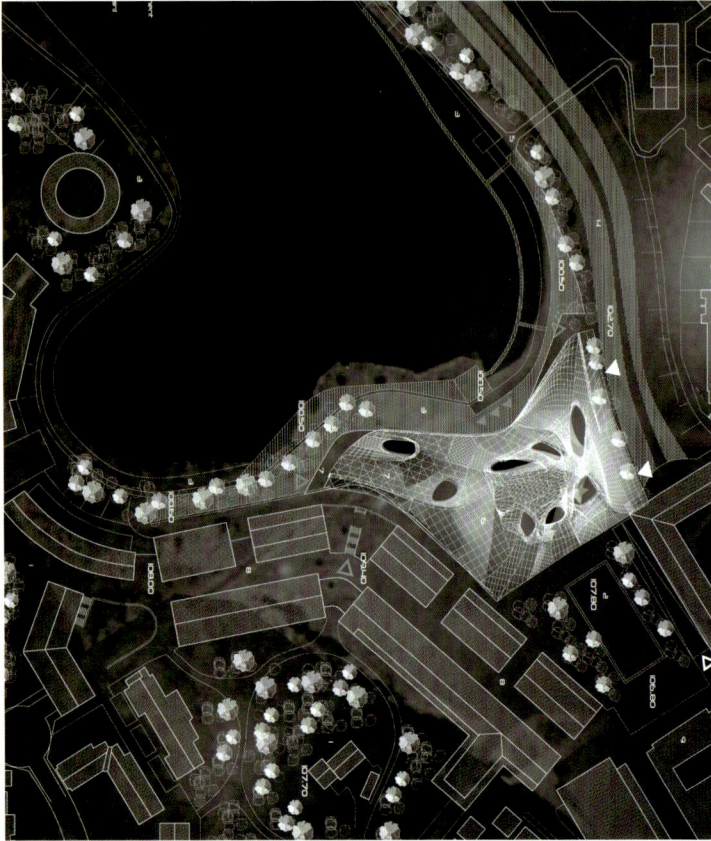

福克萨斯的设计方案通过用单一有机形式覆盖整个项目地点的方式，创造了一个"人工景观"。或者，用他们自己的话说，"环境和人造景观交织一起，互相作用；这种对地形处理方法把自然和人文两大系统综合为一体，创造出一种持续的、进取的、创新的形象。" 蜀葵颜色的无固定形状的建筑设计看起来非常像扎哈·哈迪德的卡里加利博物馆项目，添加了德国表现主义设计的卡通元素。

从某种程度上来说，这是最完美的方案设计，它使评审委员会在避免出现争议和想到实现这样优秀设计必然带来的令人头疼的建筑施工问题的同时，表现出真正的前卫思想。

Aerial view of model 模型鸟瞰图

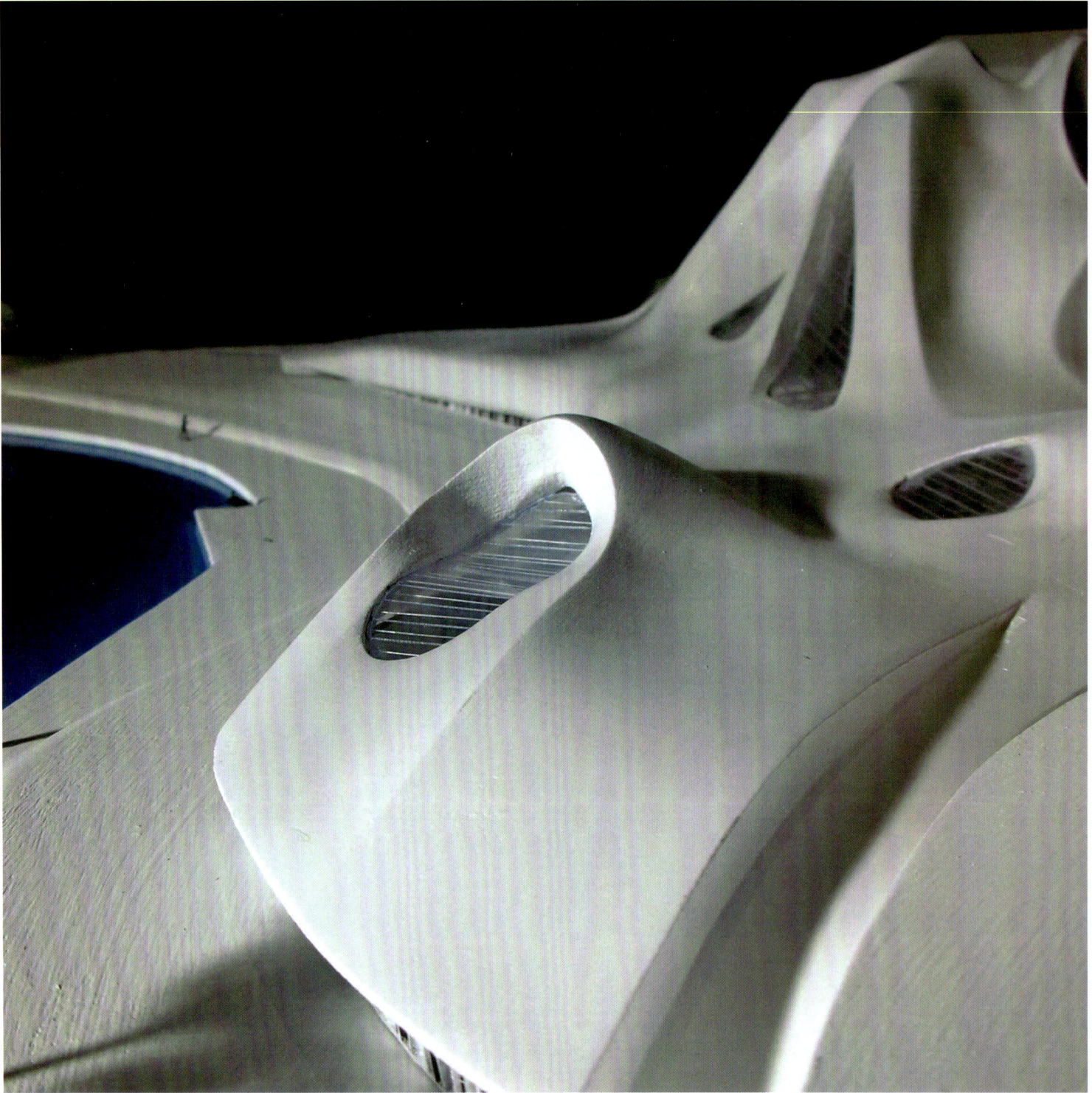

筑造学建筑事务所 参赛项目3（入围奖）

TECTONICS ARCHITECTS Ltd.
Project 3 finalist

Tectonics' response to occupying the same sliver of space is seemingly similar to the Vandehove plan, but the Londoners' two-level design interacts more with the public garden. There are fewer galleries – two main ones per floor – but they are larger and run parallel with the length of the building. The upper level is suspended over a walkway, forming an all-weather promenade; there are abundant views of the lake.

The giant arcade offers a classical rhythm that calls to mind the grandfather of all modern museums, Schinkel's Altes Museum in Berlin. In Tectonics' rendering of the promenade at night, the moonlit waterscape is like a Friedrich painting. But the real power of the design is its vigorous massing of large concrete boxes, with thin columns, and sculptural skylight.

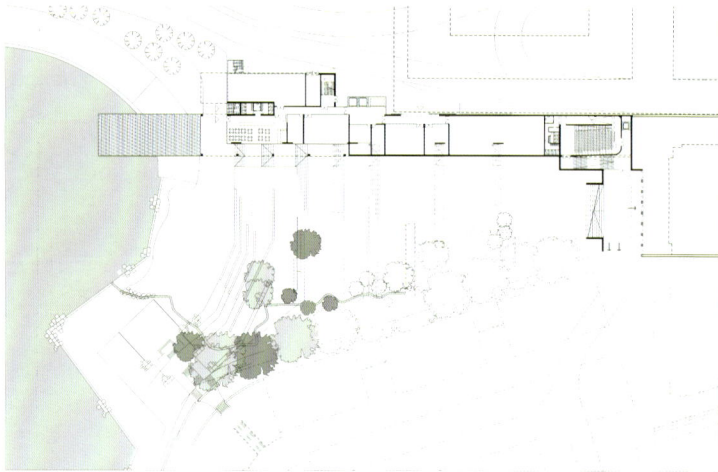

Floor plan 平面图

筑造学建筑事务所设计的相同的条状空间结构与范登霍夫的相似，但这个伦敦的建筑事务所的两层建筑的设计与公共花园的互动更多。方案中的画廊更少一些——每一层有两个主要的画廊——但是它们要大得多，与整个建筑的长度平行。二楼的下面是一个人行道，形成了一个全天候长廊，湖区的景观一览无遗。

这个大型拱廊造型典雅，使人想起柏林的老博物馆，这家博物馆是所有现代博物馆的始祖。在筑造学建筑事务所提供的夜晚长廊的效果图里，月光下的水景犹如弗里德里希的绘画作品。但是这一设计方案的真正魅力是一些充满活力的大型混凝土建成的盒状结构,柱子细细的，天窗经过精心雕刻。

Aerial view of model 模型鸟瞰图

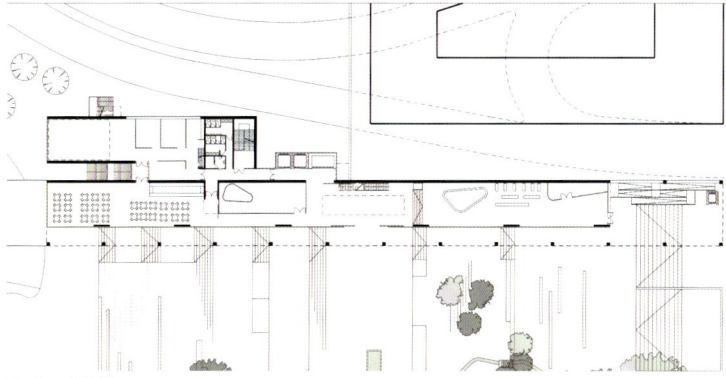

Section 剖面图

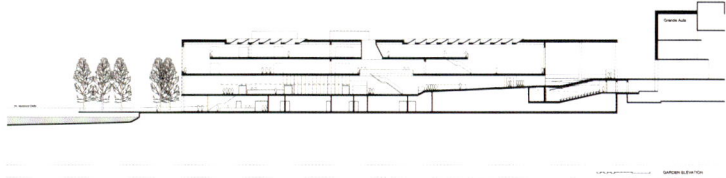

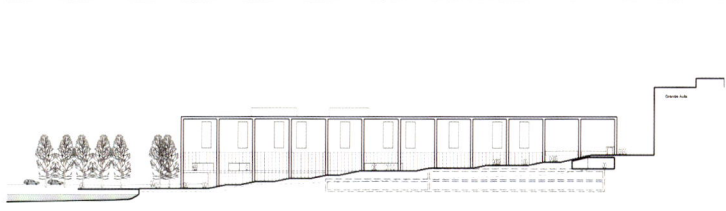

Elevation 立面图

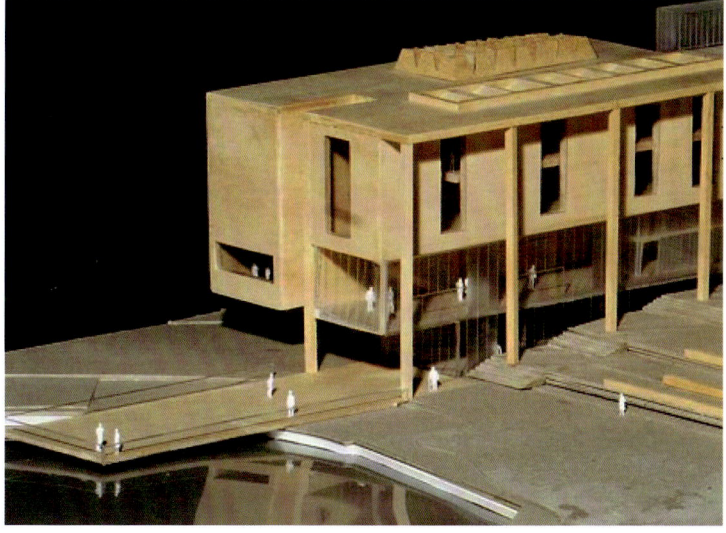

查尔斯·范登霍夫联合建筑事务所
参赛项目4（入围奖）

Charles Vandenhove et Associes
Project 4 finalist

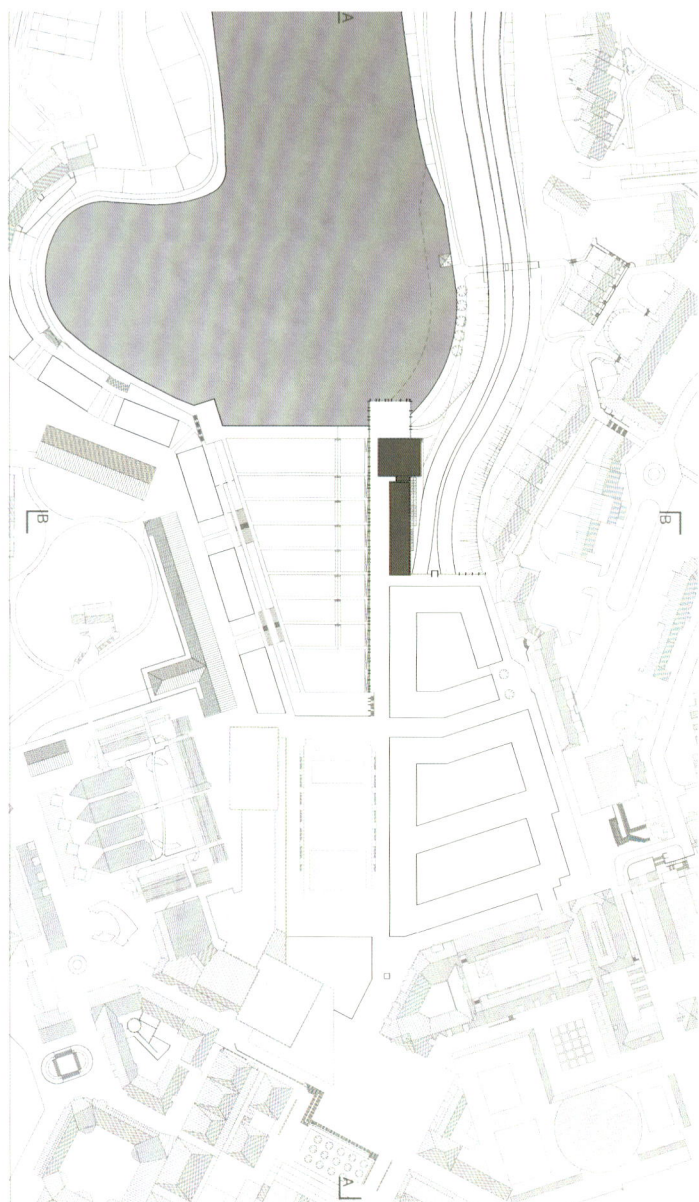

The Charles Vandenhove's design is severely Modern, with a prominent glass cube and a long arcade; but their scheme blends surprisingly well with the faux-picturesqueness of the Leon Krier-like new town. The linear Vandenhove proposal dealt with the awkward triangular-shaped site by placing the museum on a long narrow slice of land to one side while leaving a large formal terraced garden between the lake and the university.

The glass box recalls Peter Zumthor's art museum in Bregenz, Austria, but here the square crystal is not freestanding, instead rising from a plinth formed by the rest of the museum. The crystal container contrasts with the main gallery space and outdoor arcade, both of which are masonry. The plinth serves as an outdoor belvedere, part of which extends out over the lake – a grand public space.

Vandenhove's rigorous formality emphasizes museum planning more than exterior flash. Its near Miesian simplicity forms a neutral backdrop to the art.

查尔斯·范登霍夫联合建筑事务所的设计是现代的，有一个巨大突出的玻璃立方体和一条长长的拱廊，但是方案与风景如画的周围环境惊人地融为一体。范登霍夫设计的线形结构通过把博物馆放置在一大片狭长土地之上，在湖区与大学之间留出一个大型台地花园，解决了项目地点形状不规则的问题。

玻璃立方体使人想起彼得·卒姆托在奥地利的布伦根茨设计的艺术博物馆，但是这里的方形晶体不是单独立着，而是从博物馆其余部分形成的一个基座上向上腾起的，与主要的画廊空间和室外走道形成对比，它们都是砖石结构砌筑而成的。这个基座可以作为户外观景楼使用，一部分延伸到湖区，形成一个开阔的公共空间。

范登霍夫的严谨的设计形式强调更多的是博物馆的规划，而不是浮华的外部装饰。这种近似密斯简洁大方的建筑式样构成一个艺术的中性背景。

View from lake 从湖面看博物馆

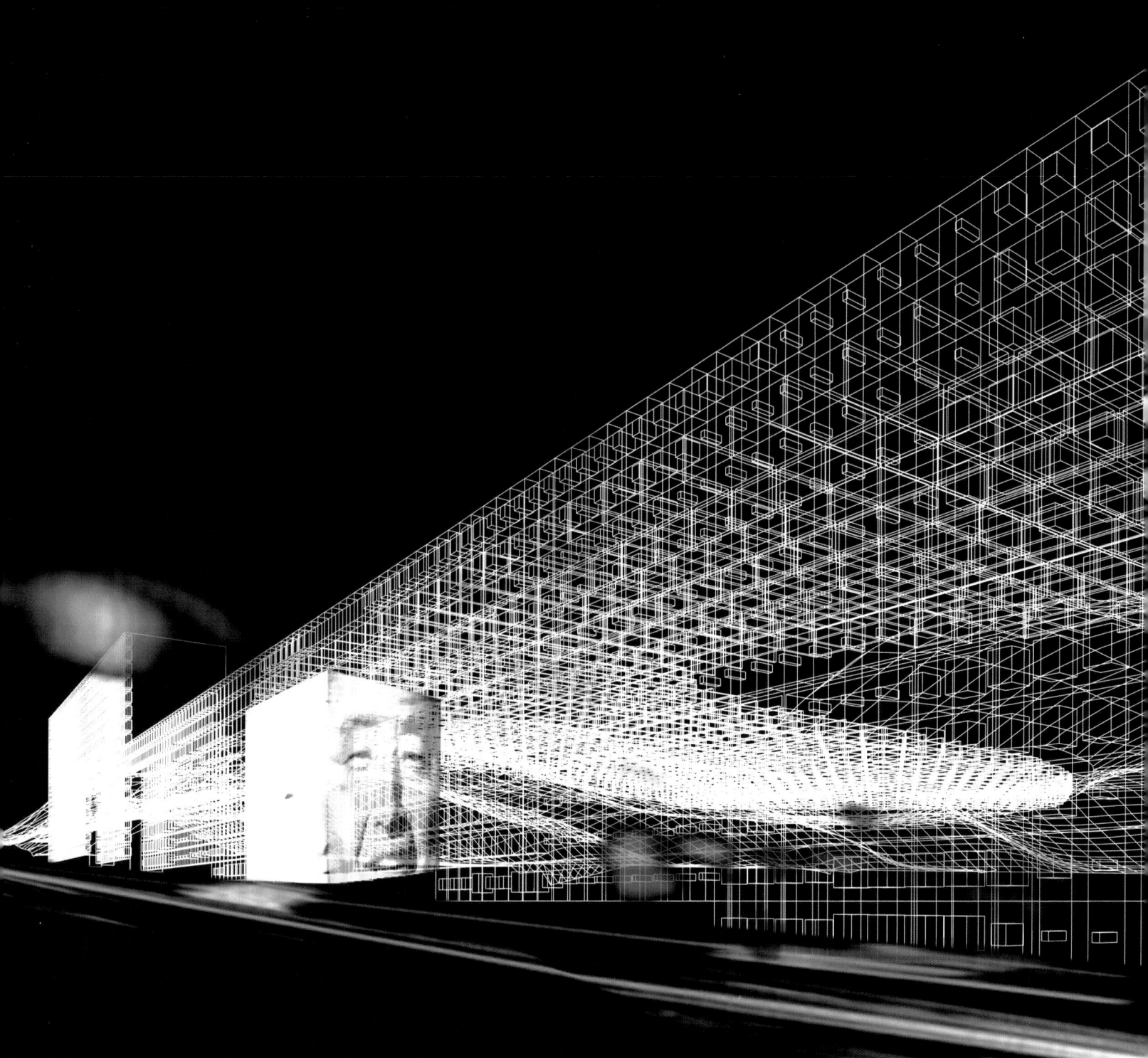

Nam June Paik Museum Competition

白南准博物馆竞赛

竞赛背景 Competition Background

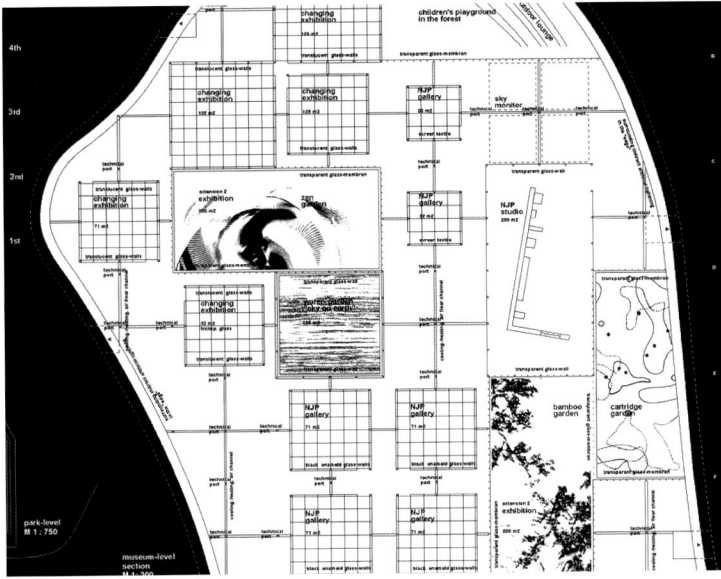

If there ever was a museum of discovery, this could well be it. The winning competition design for a new museum honoring the works of Korean artist, Nam June Paik, is all about approaching culture with a large dose of adventure added. The design "Matrix" seeks to enable museum visitors to create their own program. Although there are plenty of hints, one can pick and choose — and then come back for more.

如果世上有一家发现博物馆，那么就是这一家。在纪念韩国艺术家白南准的博物馆竞赛中，获胜的设计完全让人们通过一系列的探索来接近文化。设计"母体"试图让博物馆的参观者创造属于自己的规划。尽管有许多线索，一个人一次只可以选择一种，然后回到起点重新开始。

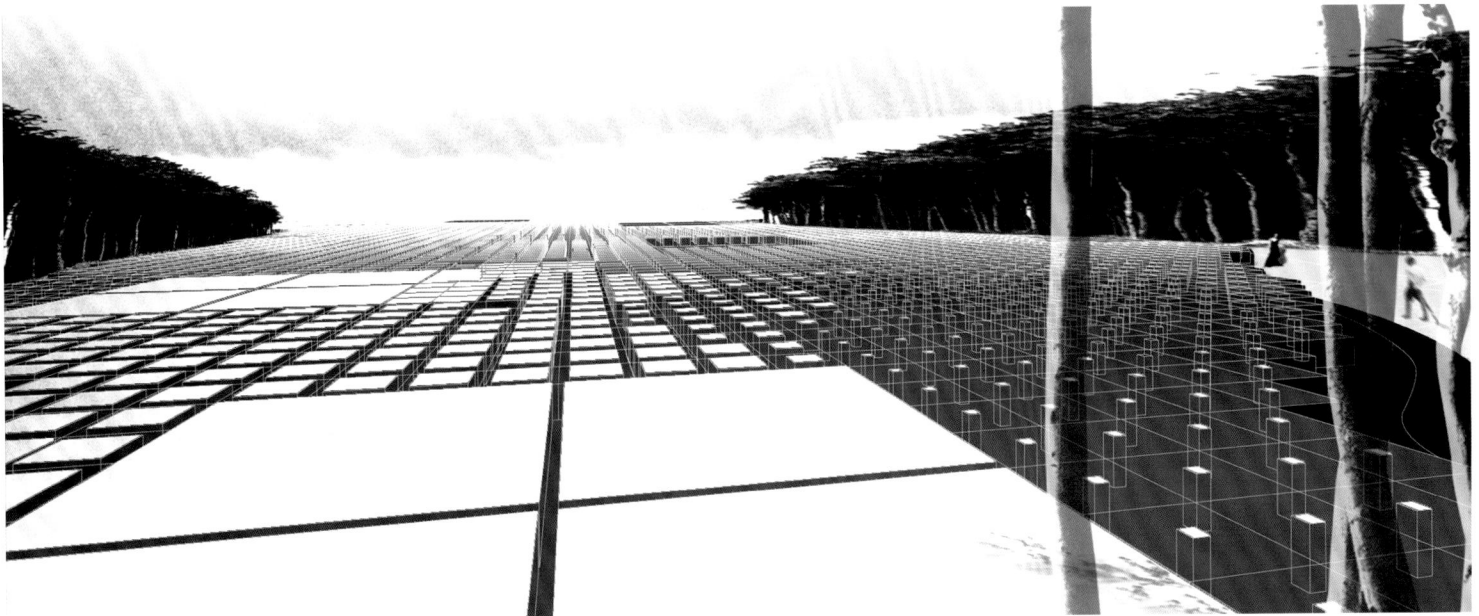

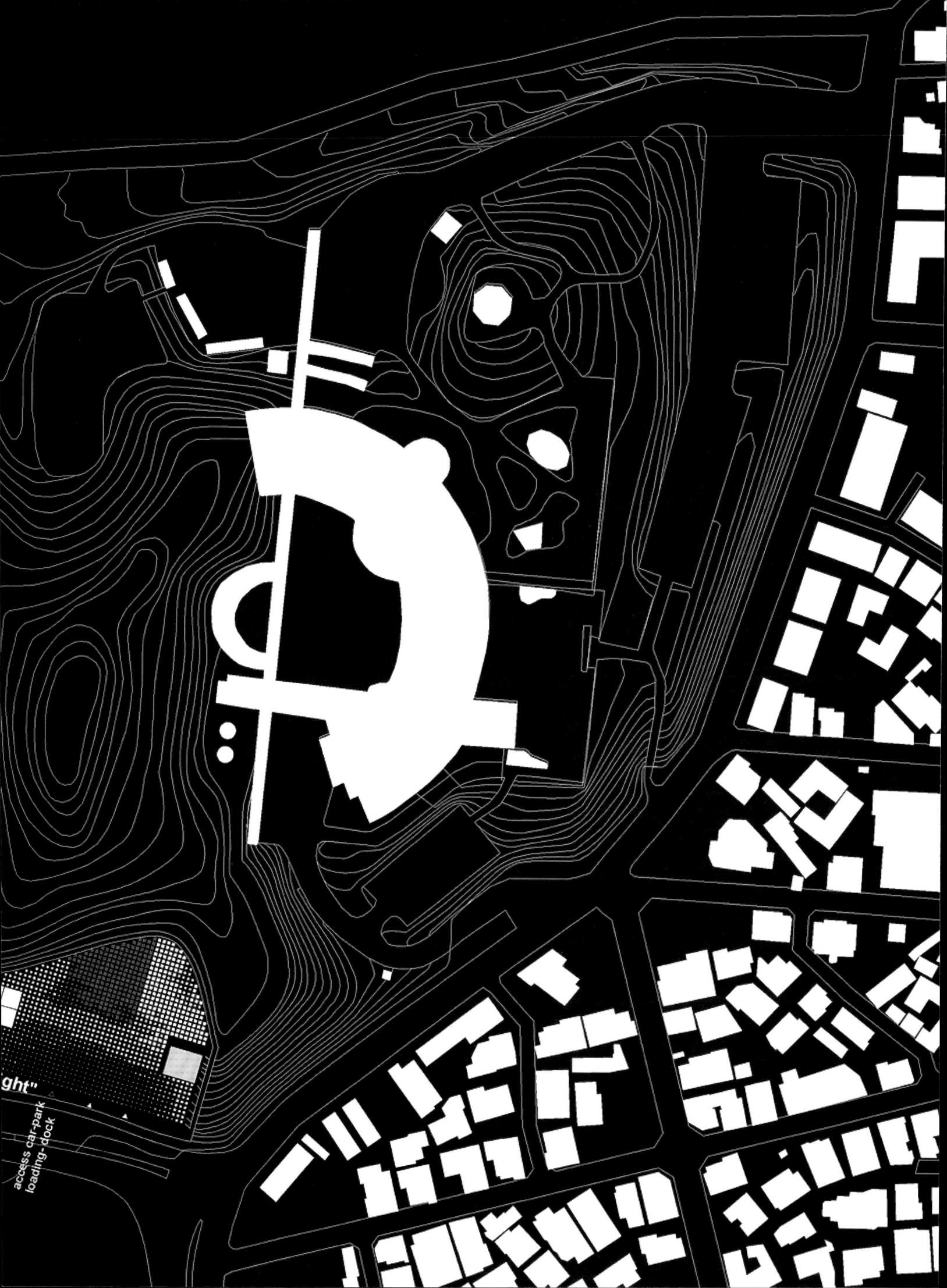

access car-park
loading-dock
ght"

专家点评 **Remarks**

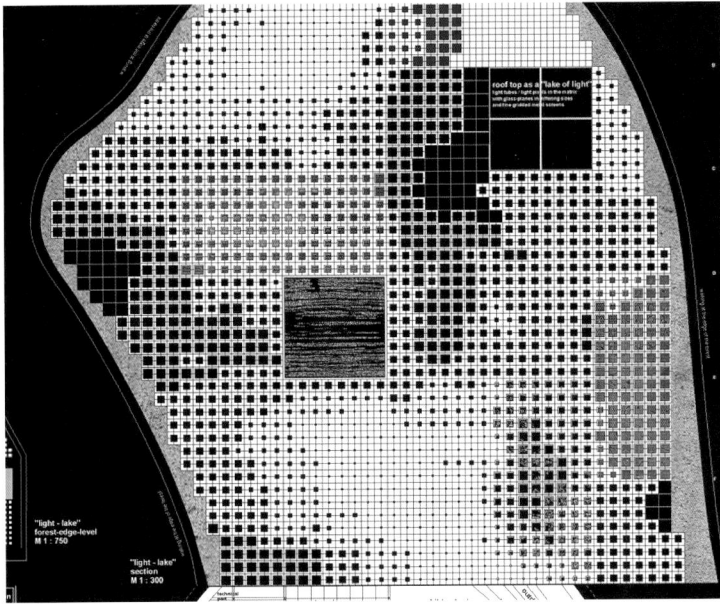

The winning design, in contrast to the runner-ups, is more about integrating the museum into the hilly landscape, rather than simply using it as a contributing factor — to park the various components.

与其他入围的设计相比，获胜的设计更注重将博物馆融入山景里，而不是仅仅将它作为一个工具，摆放各种元素。

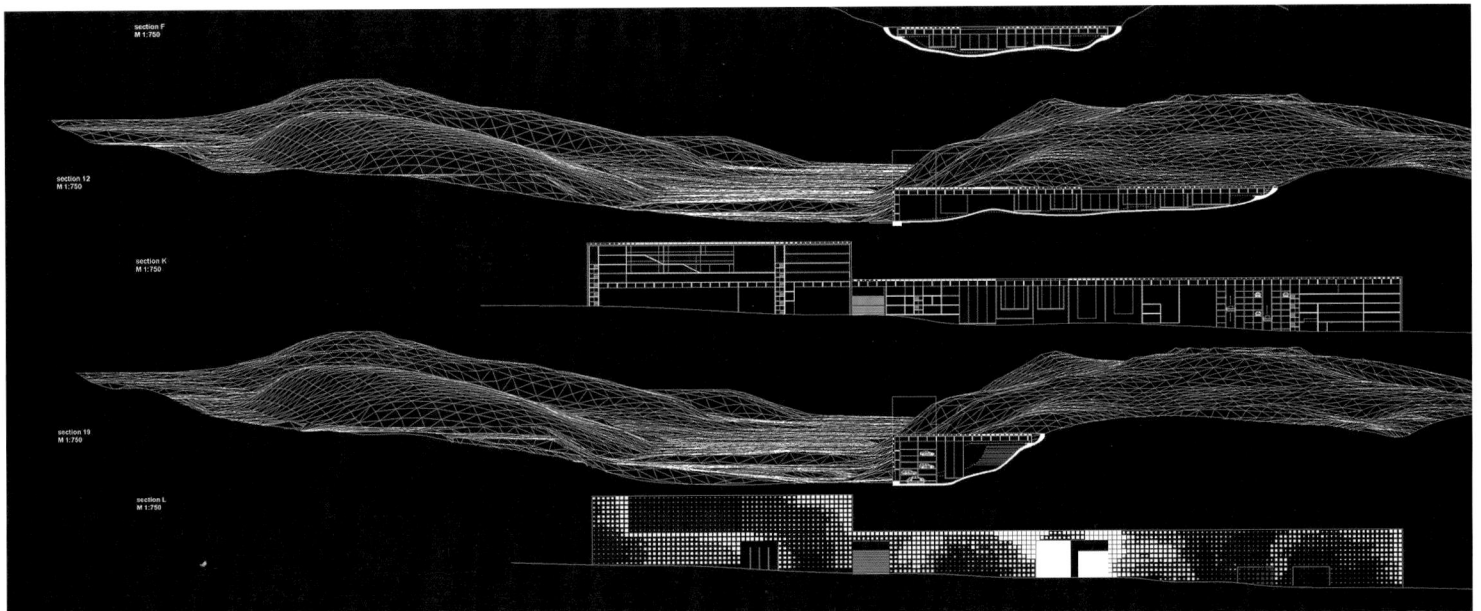

科尔斯顿·施梅尔 参赛项目1（一等奖）

Kirsten Schemel Project 1
the first prize

Floor plan 平面图

The appeal of Kirsten Schemel's scheme was based to a great extent on its flexibility. It suggested anything but rigidity, and in essence represented a "shell," rather than a finished, highly defined space. According to Schemel, "We don't want a kind of hierarchical linear space, we imagine a matrix with spatial complexity.
That structure is the search for a form of the pleasure of an art experience." She visualizes this structure as a continuation of the landscape — tucked into a depression between two hills. The use of interior gardens reinforces this impression. Although the roof consists of a large, continuous plane, the interior is broken up both visually and physically by the topographical descent created by the landscape.
Moving from level to level, rather than room to room, is the main organizational factor while wandering through the museum. Creating different levels also can break up the interior so that it does not resemble a giant convention hall. Visually, one is somehow reminded of Mies' National Gallery in Berlin, but instead of being situated on a platform, it is inserted into a landscape in a more fragmentary manner.

科尔斯顿·施梅尔设计的诱人之处一大部分在于它的弹性。它一点都不僵硬，从本质上看它只是一个"外壳"，而不是一个完整的、精雕细琢的空间。施梅尔说："我们不需要一个层次分明的线性空间，我们设计一个母体，里面空间复杂。"她将博物馆的外观设计成风景的延续，夹在两座山之间的凹地里。室内花园的运用加深了这种印象。尽管屋顶是一个大型延续的平面，室内在视觉上和实质上都被景观的地形上升分割开了。
不是从一个房间到另一个房间，而是从一层到另一层——这是参观博物馆的顺序。从视觉上来看，它和米尔斯在柏林设计的德国国家美术馆相似，但它不是修建在平地上，而是嵌在一片风景之中，"支离破碎"。

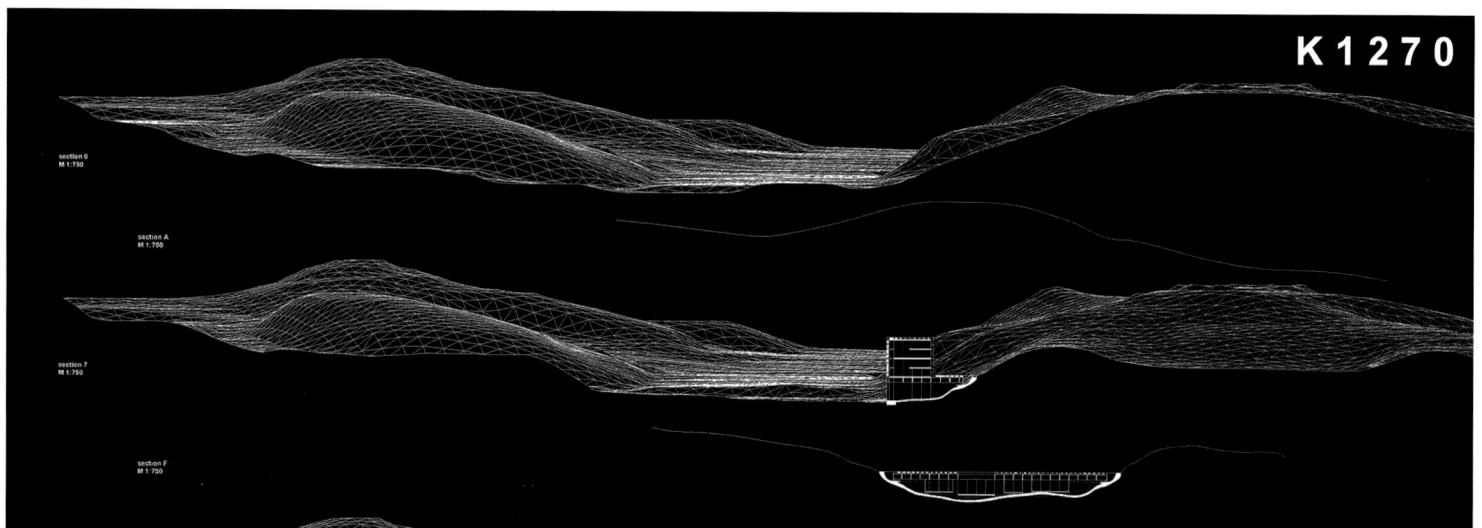

Sections 剖面图

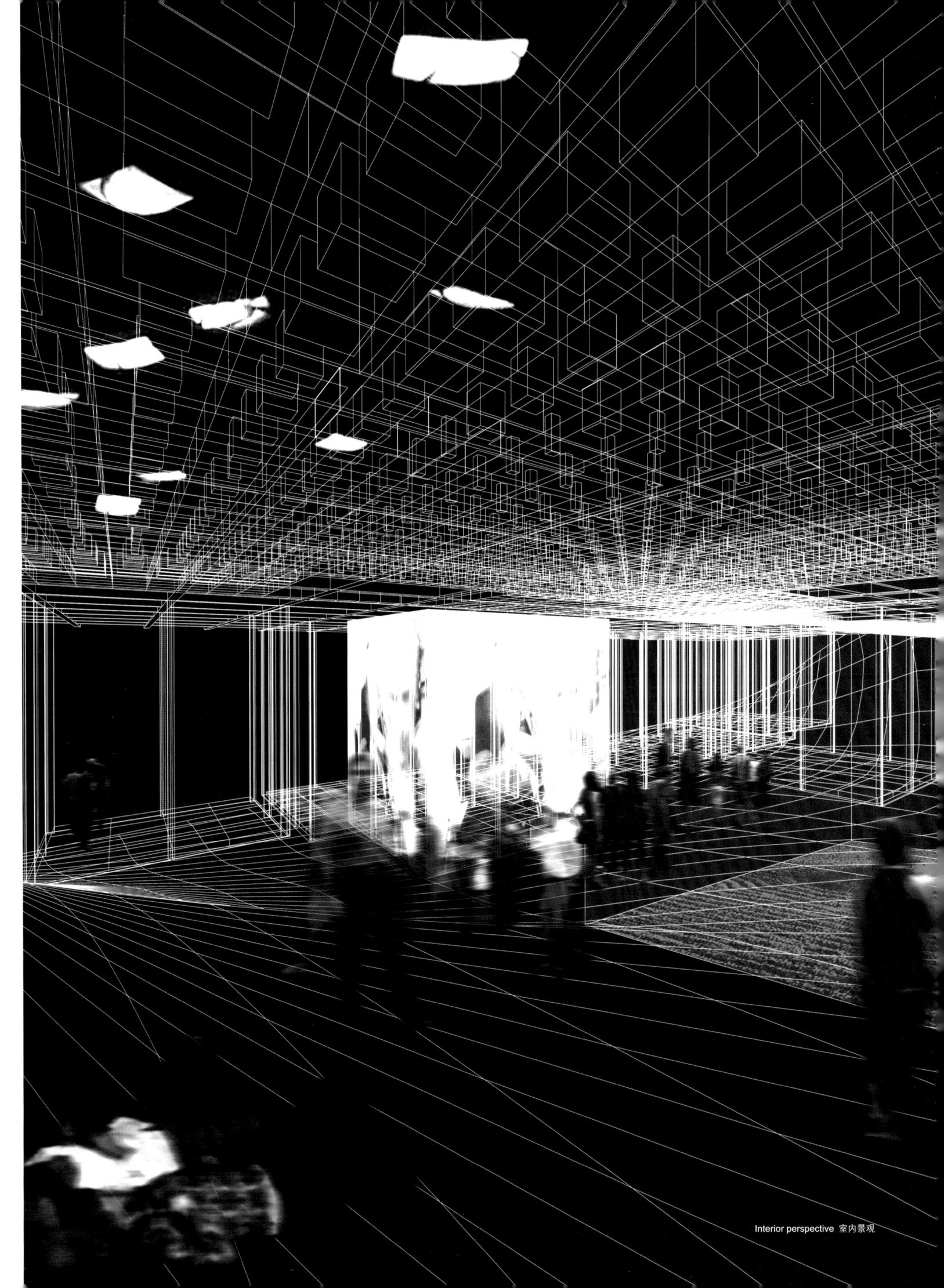

Interior perspective 室内景观

桂成宇 参赛项目2（二等奖）

Kyu Sung Woo Project 2 the second prize

The scheme of the runner-up, Kyu Sung Woo, was more typically Korean in that it was a single building with two platforms imbedded in the landscape. A sculpture garden was to be located on the side away from the road. This design was a favorite of the Korean members of the jury, who no doubt recognized that it might well be a work of a Korean architect, even though the envelopes had not yet been opened.

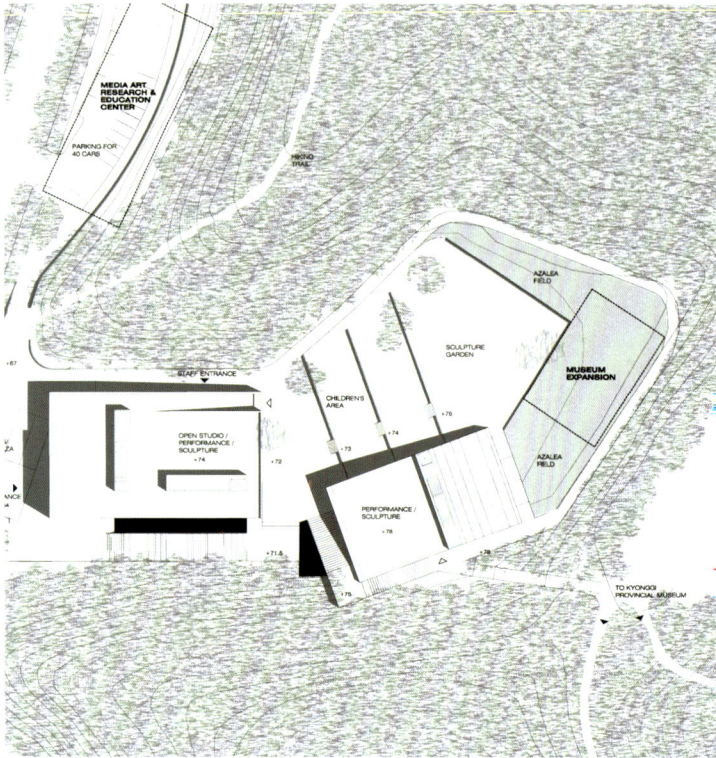

Site plan 总设计图

二等奖获得者桂成宇的作品更具韩国本土特色。他设计了一座有两个平台的建筑，原路面的一侧上有一个雕塑花园。这个设计深受各位韩国评委的欢迎，他们即使在不知情的情况下，也能确认这一设计出自一个韩国设计师之手。

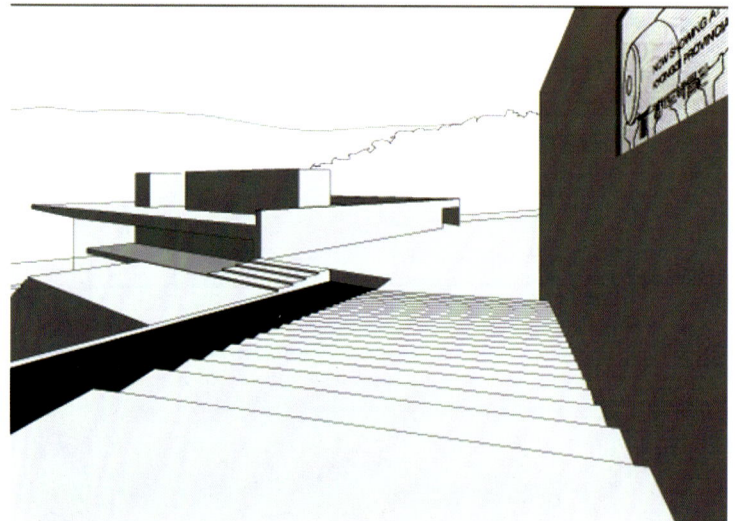

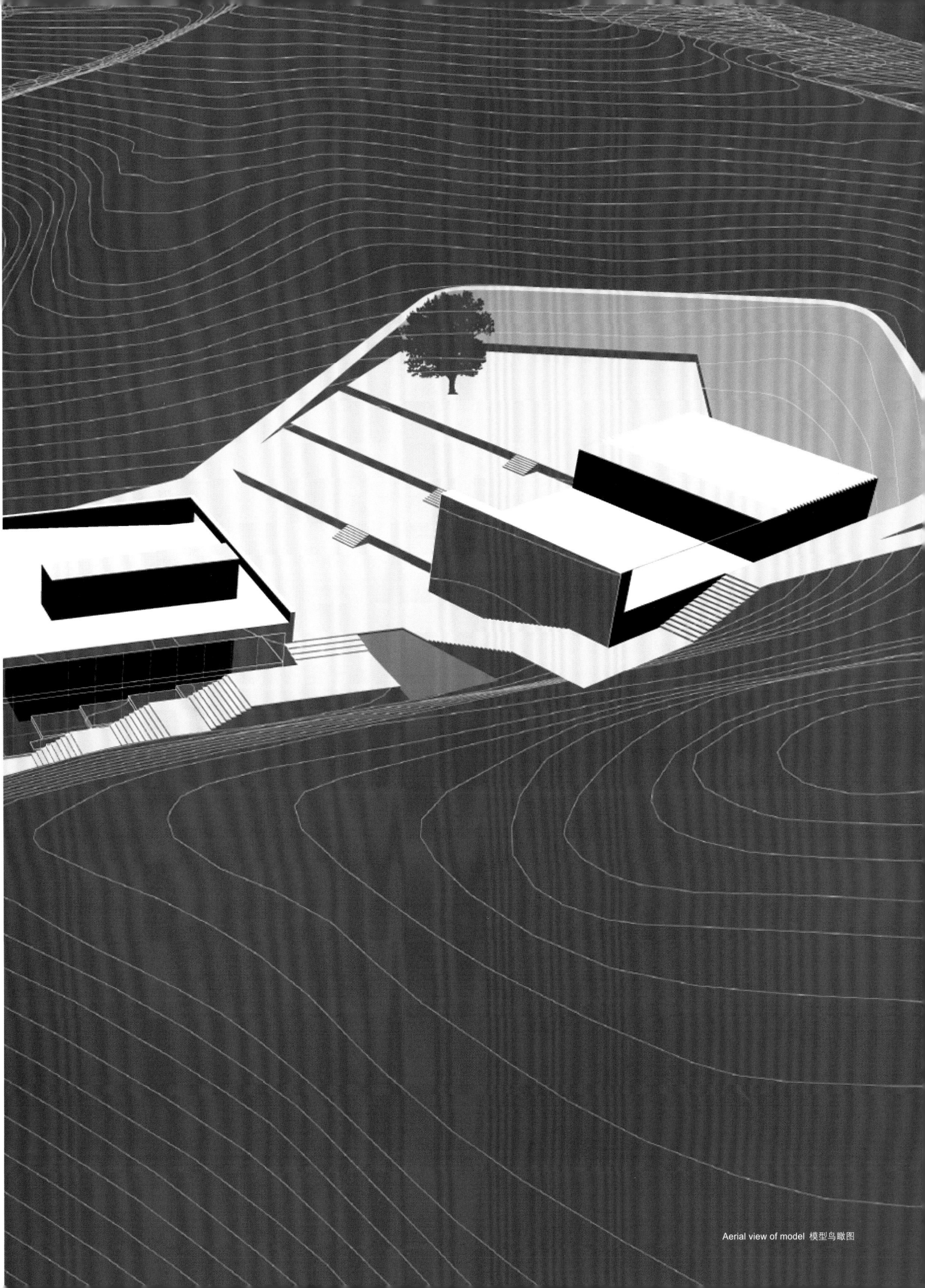

Aerial view of model 模型鸟瞰图

冈部宪明 参赛项目3（三等奖）

Noriaki Okabe Project 3
the third prize

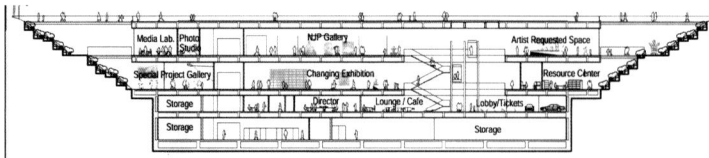

Longitudinal section 纵向剖面图

Third place went to the Japanese architect, Noriaki Okabe, who used a bridge to bring attention to the work of the artist. Here we have the exhibit areas suspended over the valley, a reasonable assumption, but not totally convincing to the jury. As this project, eventually covering an area of 9,000 ㎡, is to built in phases, one might wonder how a bridge as a flexible building platform would work in light of this program.

三等奖颁给了来自日本的设计师冈部宪明。他用一座桥来吸引人们对白南准作品的注意力。展览区域悬在一个山谷之上，这是一个合理的设想，却没有得到评审委员会的认同。由于这个项目将占地9000平方米，要分阶段进行建设，人们怀疑如何在这个计划中插入一座作为建筑伸缩平台的桥梁。

Perspective 全景图

Axonometric view of model 模型轴测图

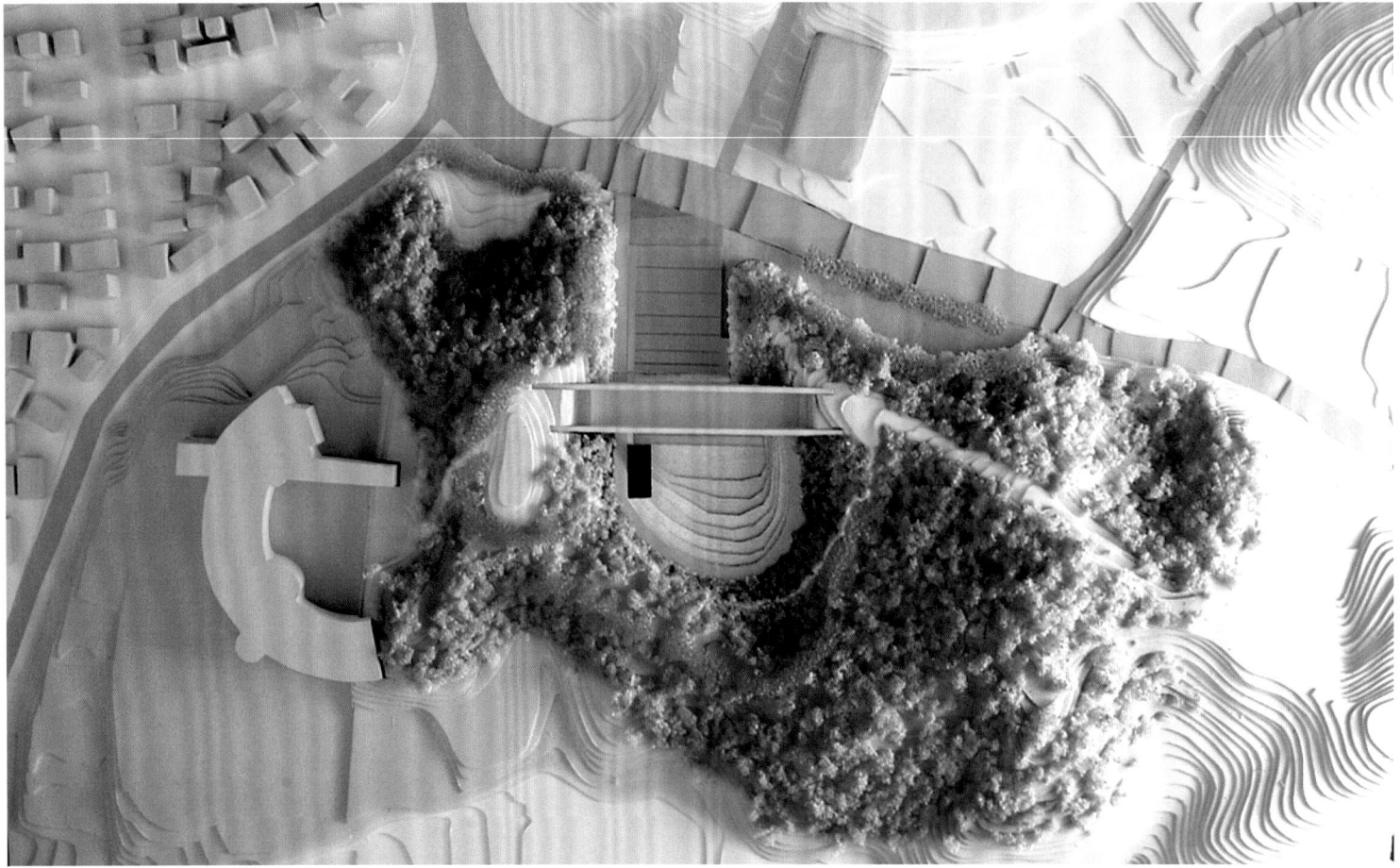

Aerial view of site 场地鸟瞰图

Aerial view of site 场地鸟瞰图

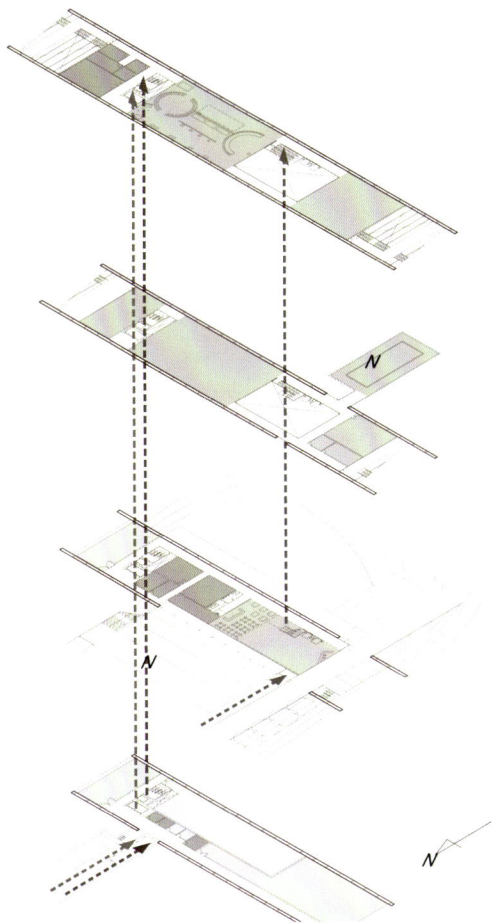

Perspective 全景图

其他参赛方案 **Other contestants' schemes**

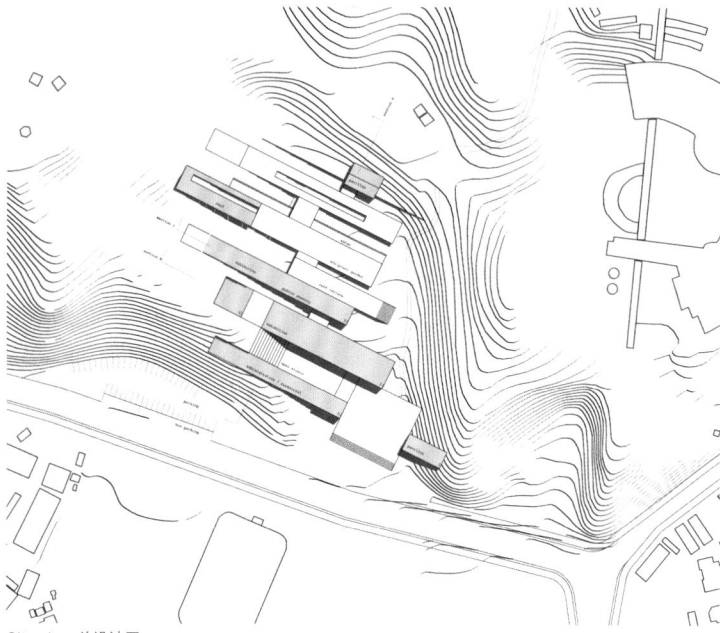

Site plan 总设计图

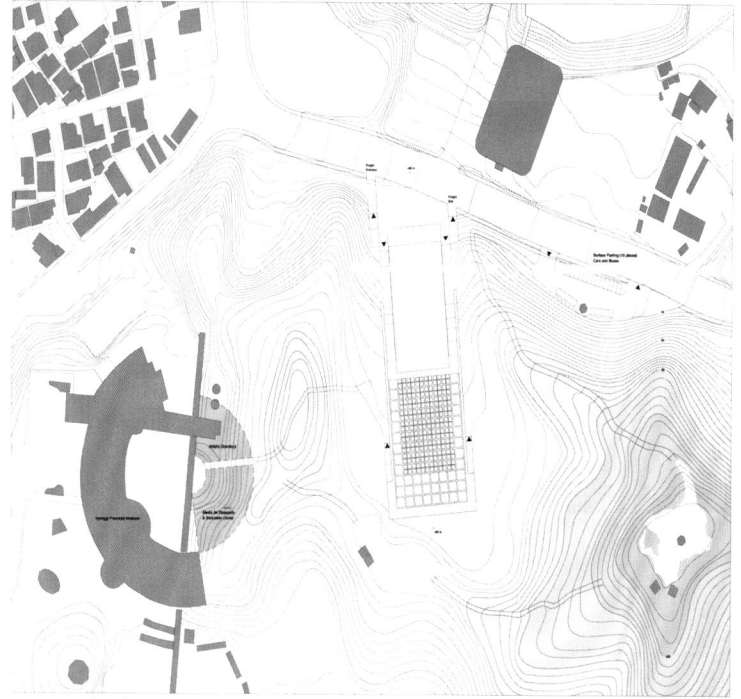

Site plan 总设计图

Section 剖面图

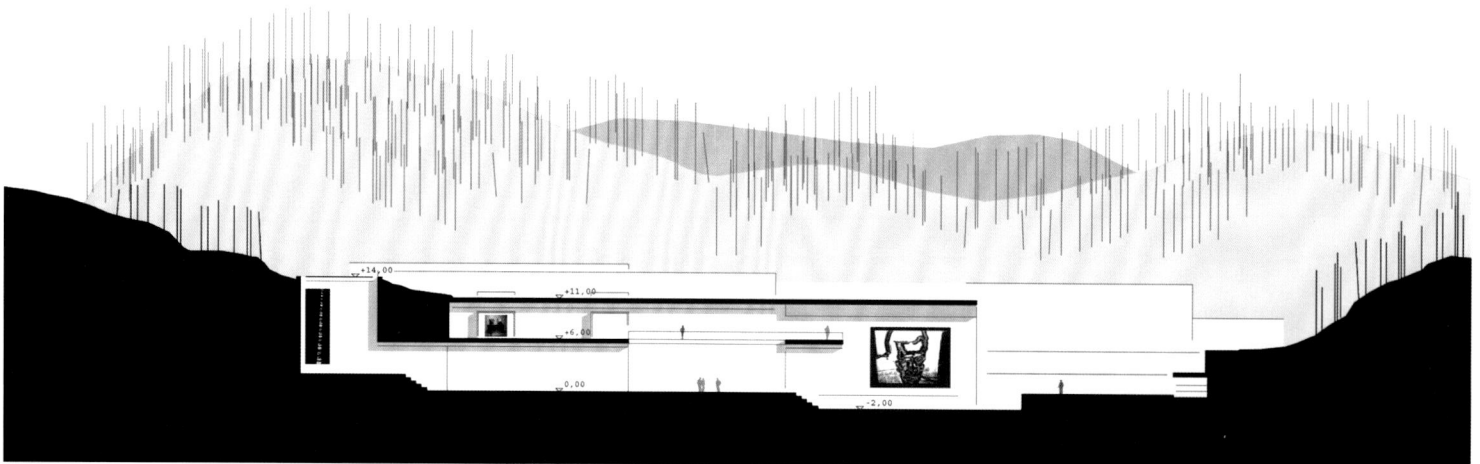

Section 剖面图

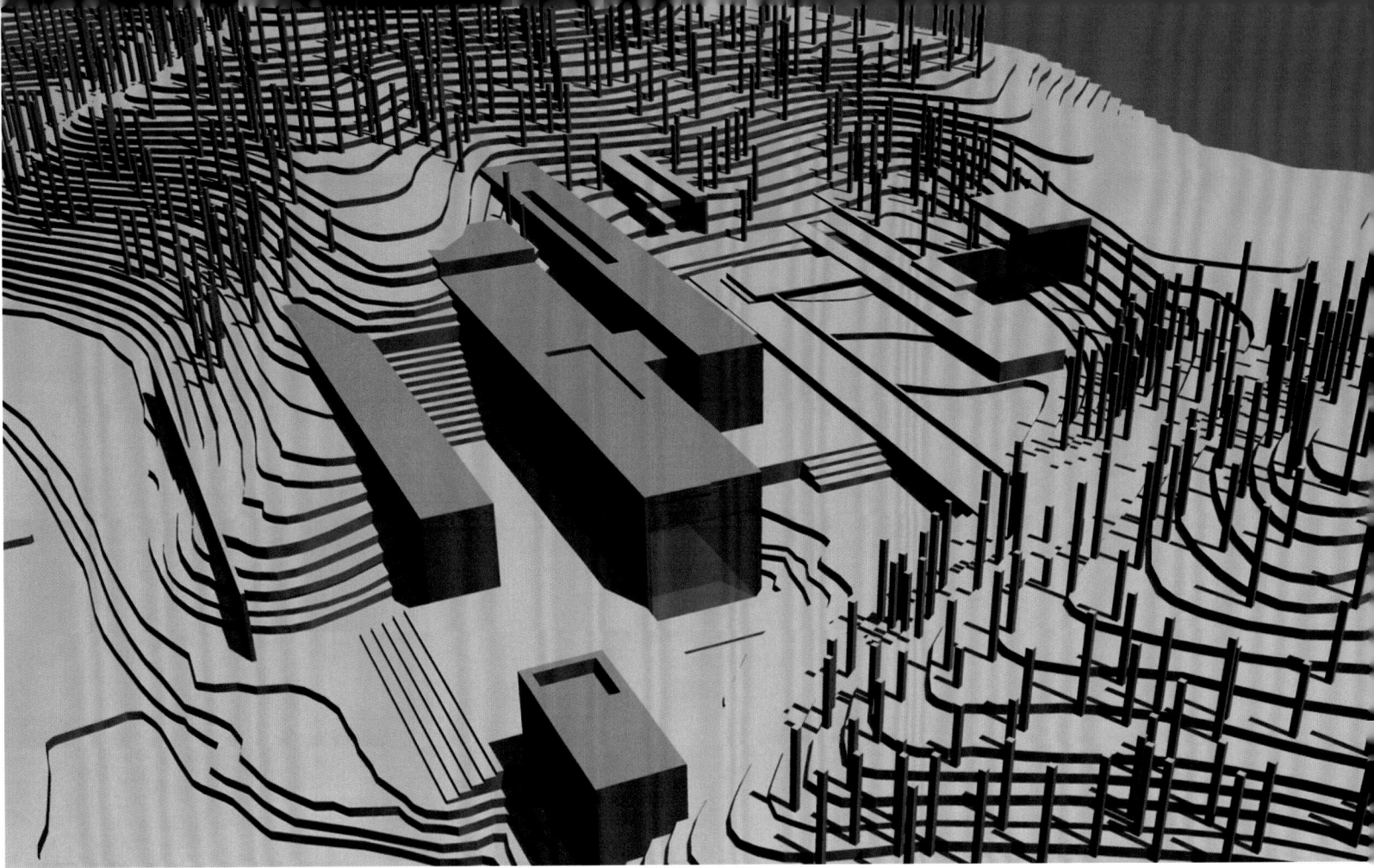

Axonometric of competition model 模型轴测图

Aerial view of model 模型鸟瞰图

San Jose State University Art Gallery Competition

圣荷西州立大学美术馆竞赛

竞赛背景 Competition Background

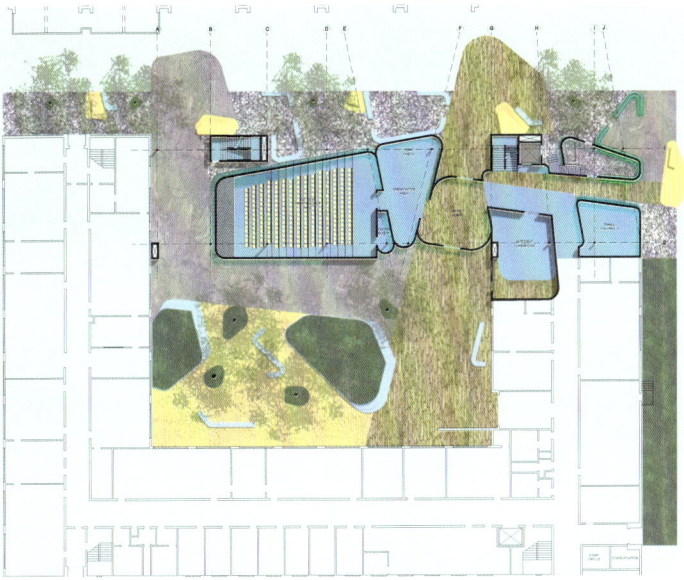

There had always been great demand for the exhibition of contemporary art and design on the urban SJSU campus, but existing facilities were not sufficiently expansive to host the numbers and kinds of displays necessary to meet this need. In mid-2002 SJSU moved a step closer to the dream of building a new Museum of Art and Design by launching an architecture competition.

圣荷西州立大学校园里总是有很多现代艺术和设计的展览，现有的设施已经无法满足这个如此数量和品目繁多的展览需求。2002年中旬，圣荷西州立大学举办了一次建筑竞赛，离梦想中的新艺术设计博物馆更近了一步。

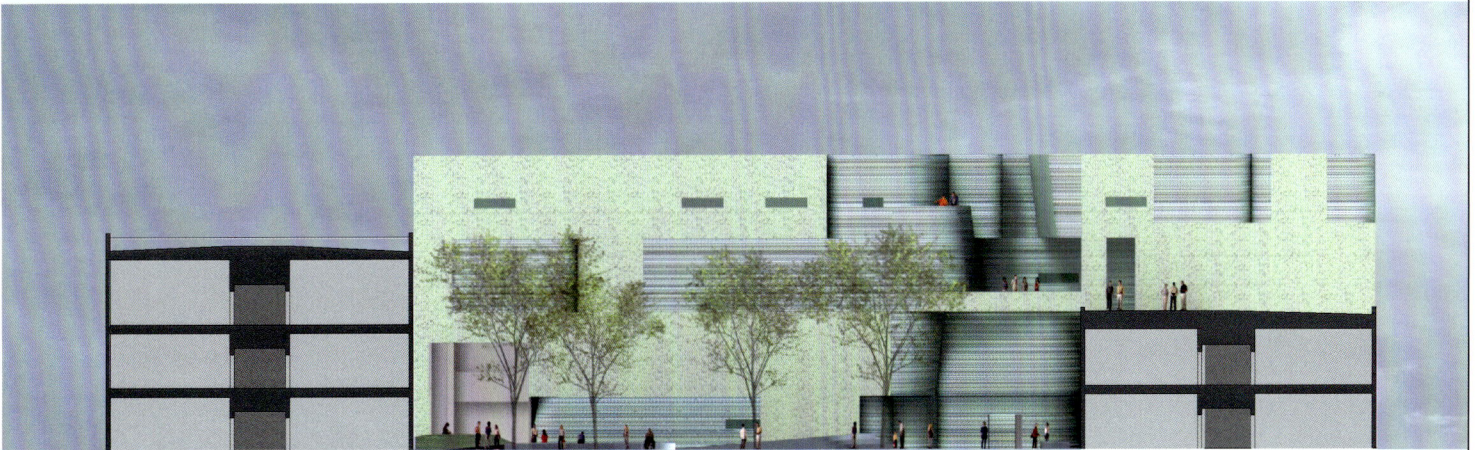

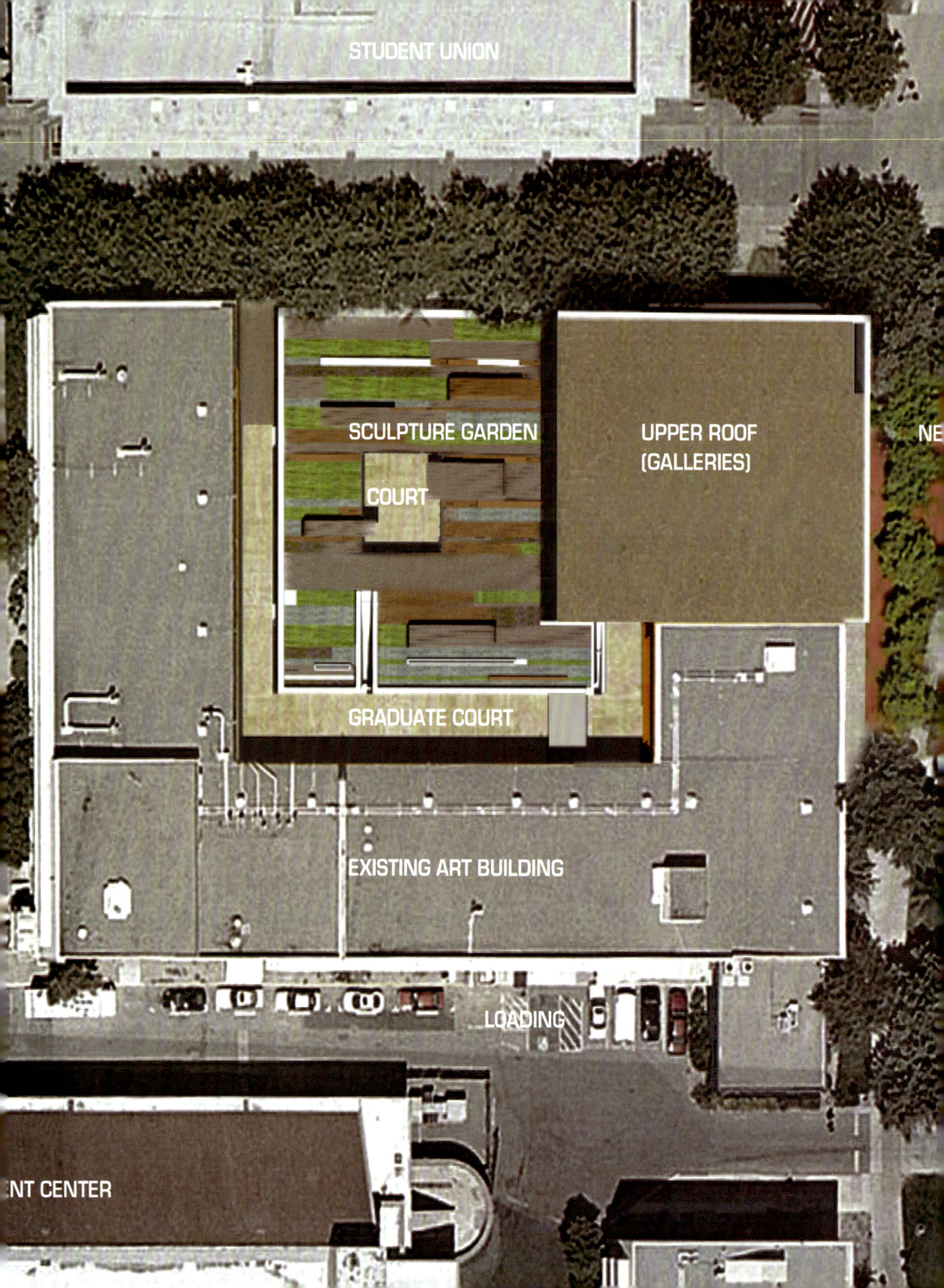

STUDENT UNION

SCULPTURE GARDEN

UPPER ROOF
(GALLERIES)

NE

COURT

GRADUATE COURT

EXISTING ART BUILDING

LOADING

NT CENTER

专家点评 **Remarks**

Planning and implementing a design competition enabled the Gallery and the University to simultaneously achieve several diverse goals that simply hiring an architect would not.

It provided significant, international visibility for the project, gave people an opportunity to review innovative ideas from various design teams.

It afforded SJSU students a concrete and experiential understanding of how the architectural design process works.

It provided the university with the opportunity to consider a numerous and widely varying selection of creative responses to our spatial and programmatic needs.

与简单的聘请一位建筑师来设计新美术馆相比，策划举行一场建筑竞赛让美术馆和学校同时达成了以下几个目标：

它为项目提供了重要的国际视野，让人们有机会欣赏来自各种设计团队的创意设计。

它让圣荷西州立大学的学生对建筑设计的流程有了一个具体的了解。

它让学校针对空间和设计需求，在设计方面有了大量而宽泛的选择空间。

WW 参赛项目1（一等奖） WW Project 1 the first prize

SECTION 1

1 RESIDENCE A	5 SMALL GALLERY
2 RESIDENCE B	6 CAFE
3 MEDIUM GALLERY	7 SMALL GALLERY
4 TERRACE	8 REPLACEMENT SPACE

In a surprising upset, the small architectural team WW won the competition. Triumphing over the other 168 architectural teams, their scheme met the challenges of the site with elegance, grace, and functionality. Appropriately, WW's design projected a very sculptural image, and both the interiors and the striking exterior façade reference the kinds of shapes, forms, and balances that artists are concerned with.

The $12-15 million building, which WW has dubbed "LOOP," informs an organic, layered approach to the spaces and volumes which form the six galleries, 200-seat auditorium, café, public reception areas, and staff offices and workshops. Seeking to balance form, program, and technology, their design enhances the circulation and traffic flow within the existing building as well as between the existing building and the new museum, increasing accessibility where it is needed and providing for protection and security when necessary. It presents a hip, urban streetscape façade, enhancing the pedestrian walkway in front of the Student Union, as it simultaneously reforms and customizes the interior courtyard, improving its aesthetics as well as its functionality. It is well poised to deliver what the mission of the new museum specifies: creating a "signature," iconic center for the University as it provides a new locus for diminishing the gap between the creation and the study of works of art and design.

出人意料，不知名的WW团队获得了竞赛的胜利。他们的设计从168支设计团队提交的作品中脱颖而出，高贵典雅而又不失实用。值得称赞的是，WW的设计显示了精雕细刻的品质，内部装修和引人注目的外墙都展现了艺术家对造型、结构、平衡的考量。

WW给这座造价1200-1500万美元的美术馆命名为"环"。"环"体现了大楼有机、分层的流通结构和空间。美术馆由六个画廊、一个有200个坐席的礼堂、一个咖啡厅以及公共接待区域、工作人员办公室和工作室组成。为了达到结构、规划和技术之间的平衡，他们的设计提高了现有建筑内部以及现有建筑和新美术馆之间的流通质量，为需要的地方增添了可达性，并且提供了必要的安全保护措施。新美术馆呈现了一个如风景画一般时尚而现代的外墙，装点了学生会门前的人行道，并通过重新改造室内庭院而提升可建筑的美感和功能性。这一设计很好地权衡了新美术馆的要求：为学校创造一个标志性建筑，为削弱艺术设计的创造和研究之间的代沟提供一个新的中心点。

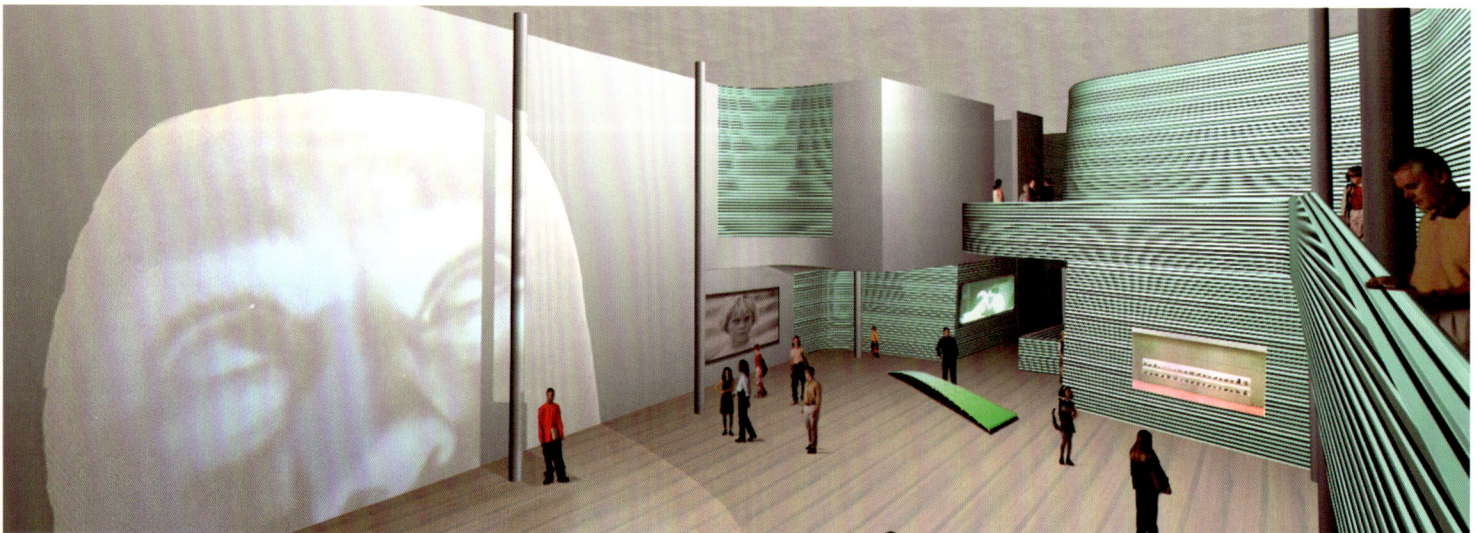

Atrium perspective 中庭全景图

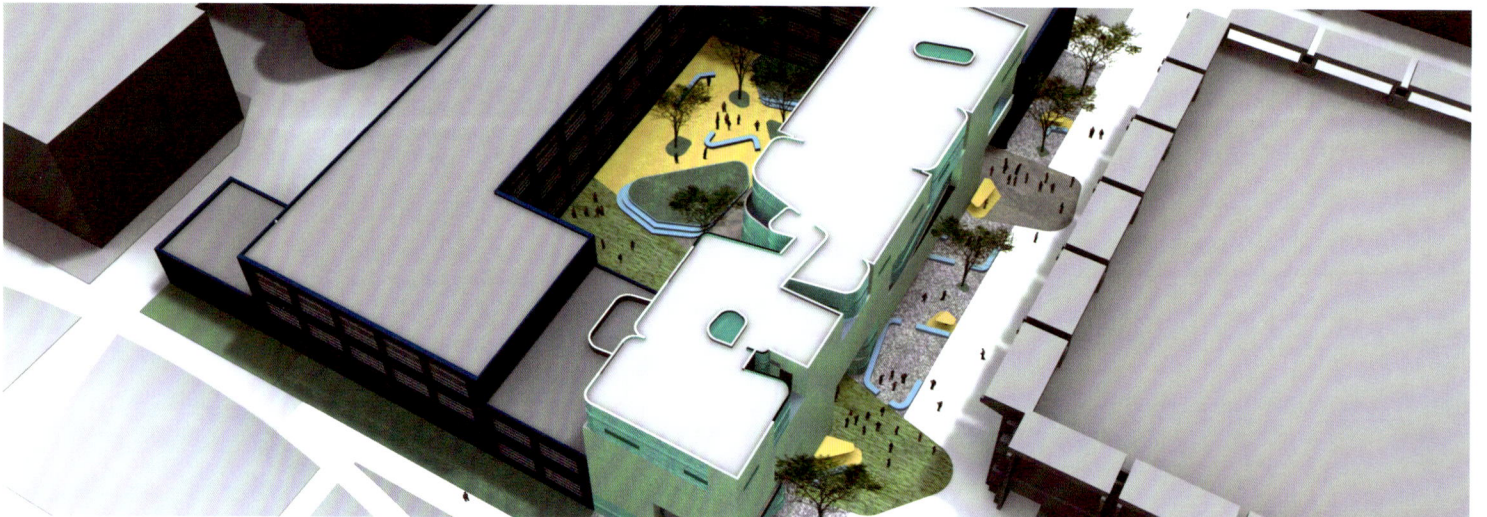

Birdseye view 鸟瞰图

Elevation 立面图

Courtyard 庭院

其他参赛方案　**Other contestants' schemes**

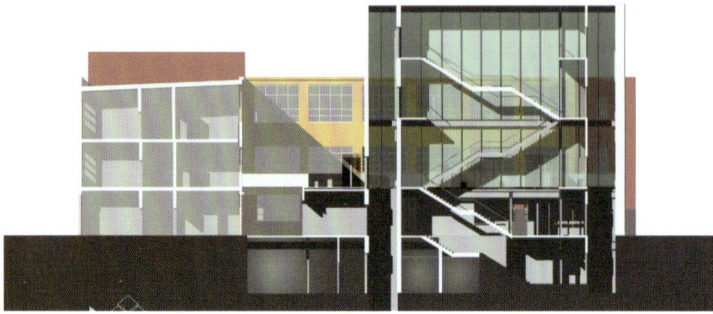

Sections 剖面图

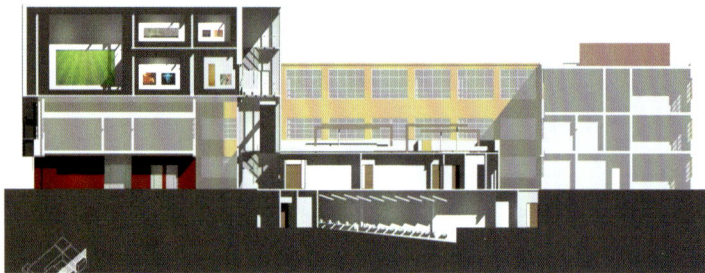

Section 剖面图

Elevation 立面图

Gallery perspective 画廊

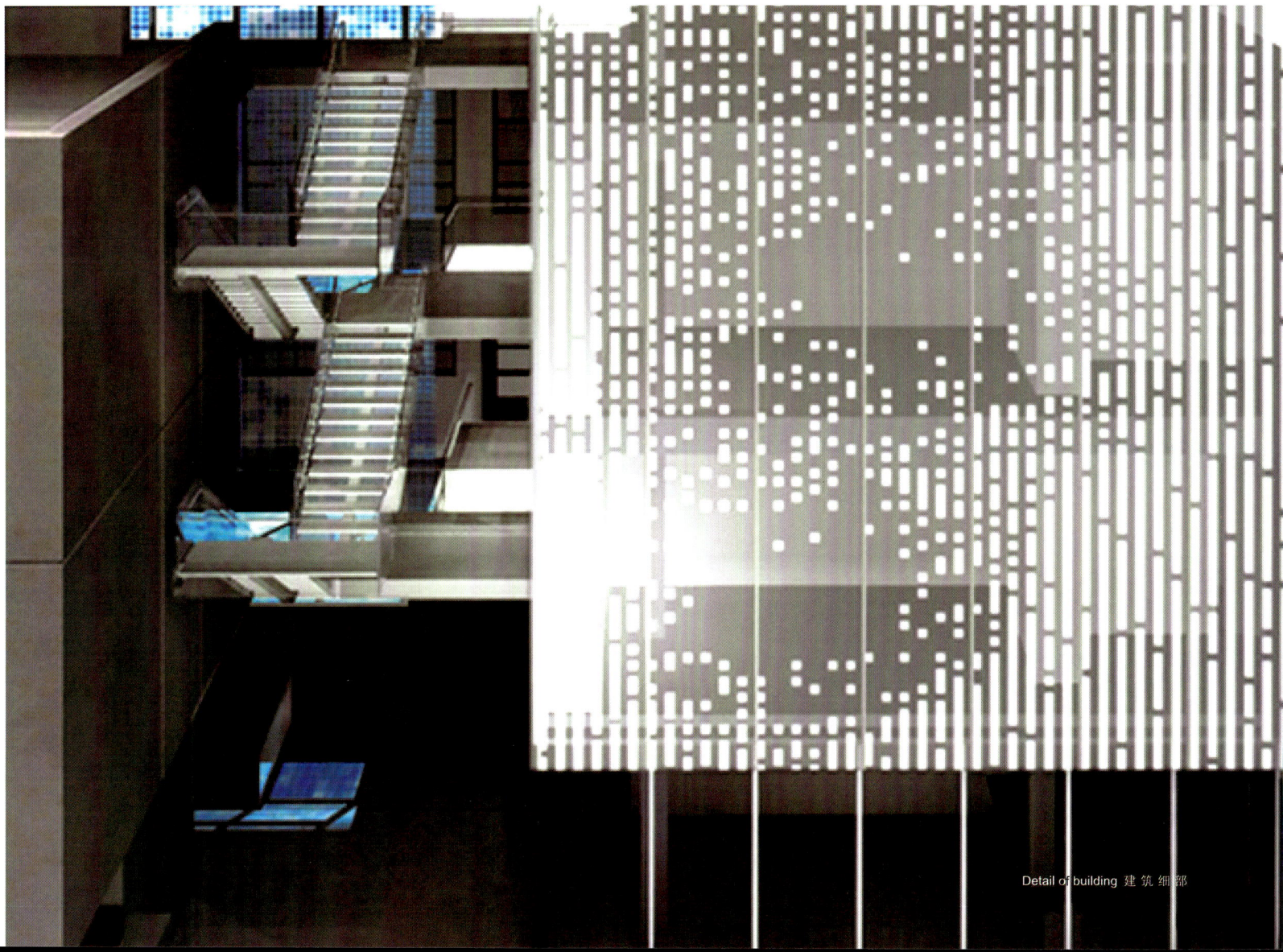

Detail of building 建筑细部

Site plan 总设计图

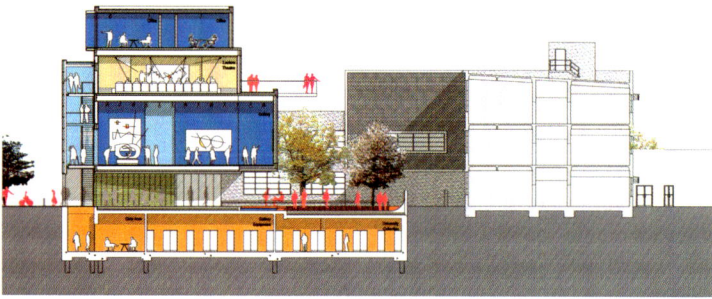

Section 剖面图

Exterior 外观图

Interior 室内图

Front elevation 正立面图

MUSEUM OF ART & DESIGN

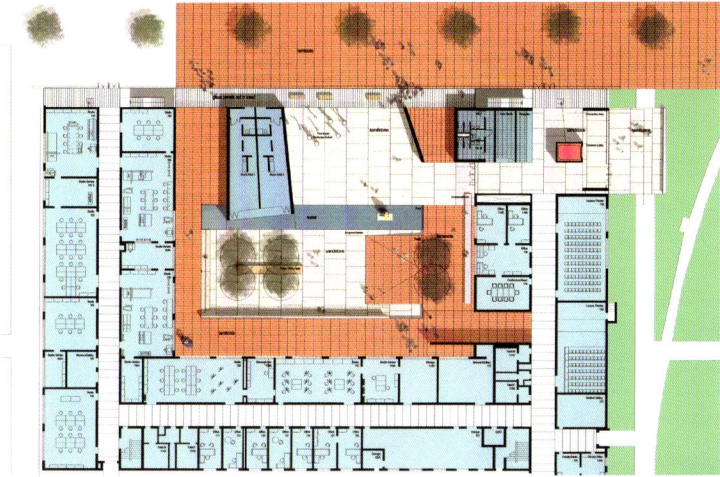

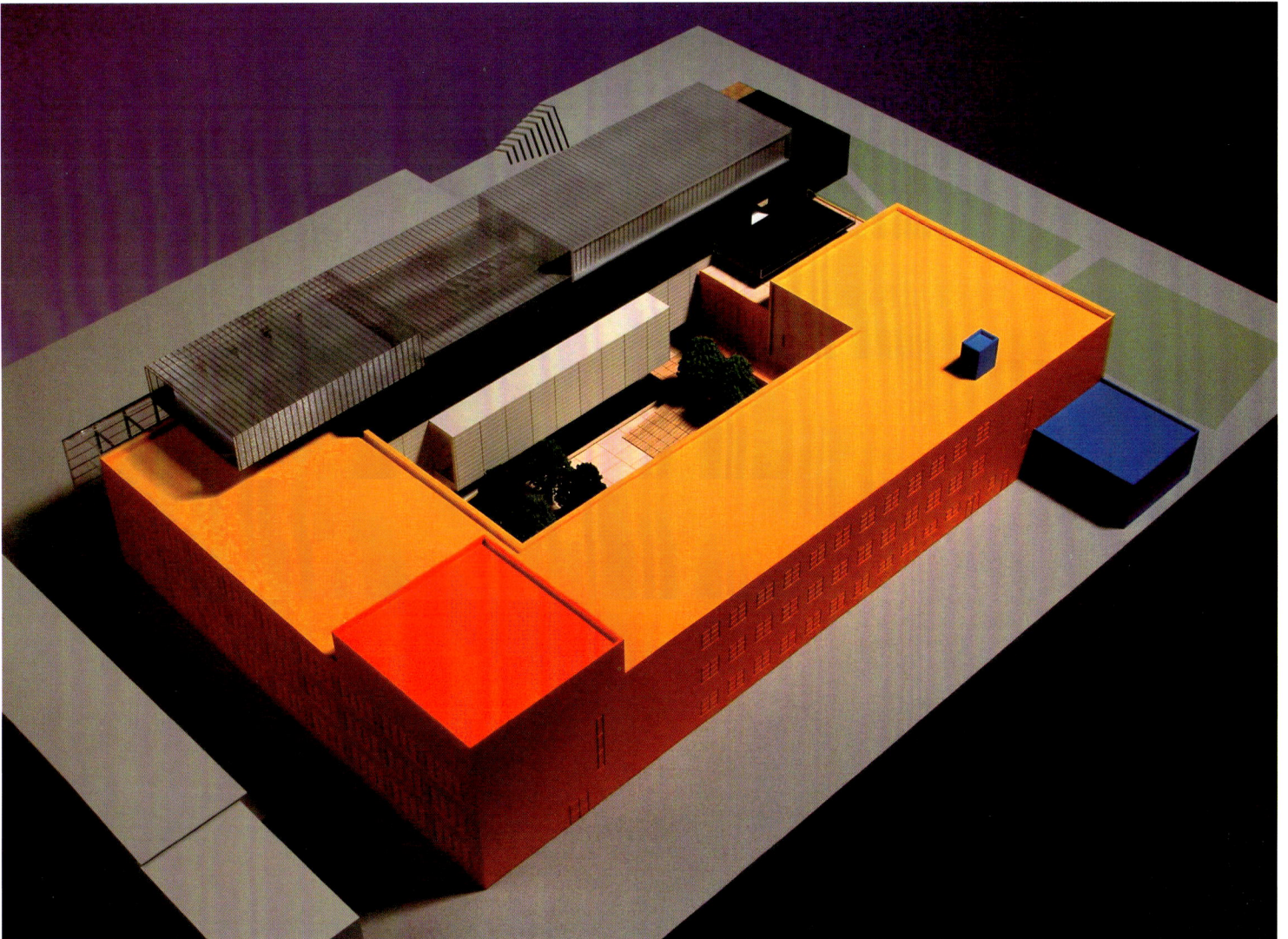

Birdseye view of model 模型鸟瞰图

The Perm Contemporary Art
Museum Competition
彼尔姆当代艺术博物馆竞赛

竞赛背景 Competition Background

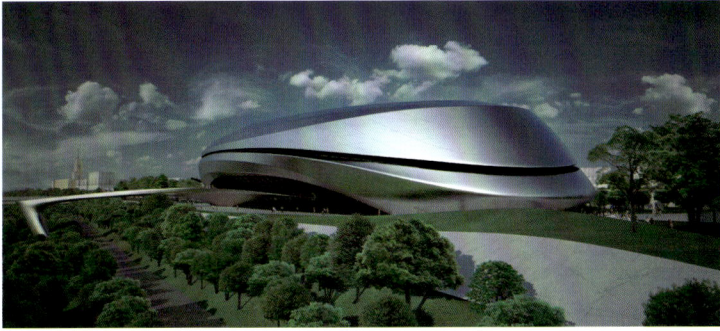

Perm Region is one of Russia's most powerful and ambitious regions. It occupies an extensive territory between Europe and Asia on the boundary between the Ural Mountains and the steppe lands. The site is located on the Kama River in Western Siberia. The new museum should provide local, avant-garde artists with a venue where more challenging modern art will have its place.

As an open competition, the event attracted entries from all over the world, with entries numbering over 300. One incentive to enter was undoubtedly the amount of prize money offered to the winners — $300000.

彼尔姆是俄罗斯最强大和有发展的地区之一，位于欧亚分界线上的乌拉尔山脉和大草原之间，占地面积广阔。项目地点位于卡玛河流域的河岸上，新博物馆应该可以提供给当地前卫艺术家一个展示更具有挑战性作品的现代艺术场所。

这次公开的竞争吸引了来自世界各地的参赛者，参赛作品超过300个。给优胜者颁发一定数量的奖金——30万美金无疑是该竞赛广泛参与的动机之一。竞赛分为两个阶段。

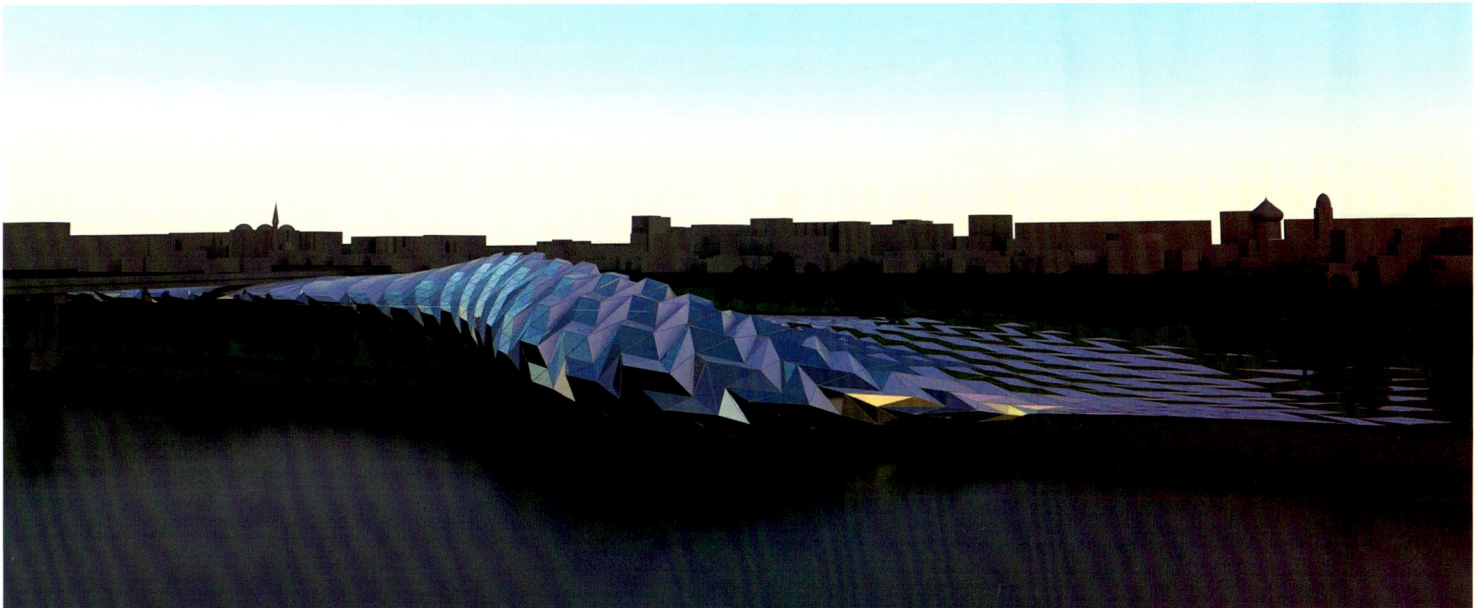

专家点评 Remarks

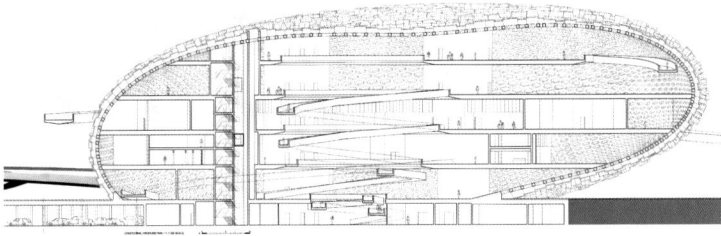

The Perm Museum, with a functional area of 16,000 sqm is a new building for Perm Art Gallery. The Perm Art Gallery is one of the most important art museums in Russia and has become the most important art centre for Ural region. This will be the first museum in Russia built to meet the needs of the 21st century. The building will be an outstanding work of contemporary architecture and Russia's most modern art museum.

This is the first attempt to design a Russian museum in accordance with 21st-century ideology and technology as a building capable of shaping the city in which it stands. The new museum serves as an urban brand, a calling card for the city and its region.

彼尔姆博物馆，功能区占16000平方米，是彼尔姆美术馆的一个新建筑。彼尔姆美术馆是俄罗斯最重要的一家艺术博物馆，已成为乌拉尔地区最重要的艺术中心。这将会是在俄罗斯建立的满足21世纪需要的第一个博物馆，该建筑将成为当代建筑的一个杰作和俄罗斯最现代的艺术博物馆。

这是第一次运用21世纪思想和技术，尝试设计一个俄罗斯博物馆，并把其作为所在城市的标志。新的博物馆作为城市品牌，是该城市及地区的一张名片。

伯纳斯克尼建筑事务所
参赛项目1（并列一等奖）

Bernaskoni Project 1
the first prize

Bernaskoni of Moscow opted to build down to the river, so that the section resembled a cascading effect, even covering the railroad tracks at the bottom riverbank to accommodate a station.

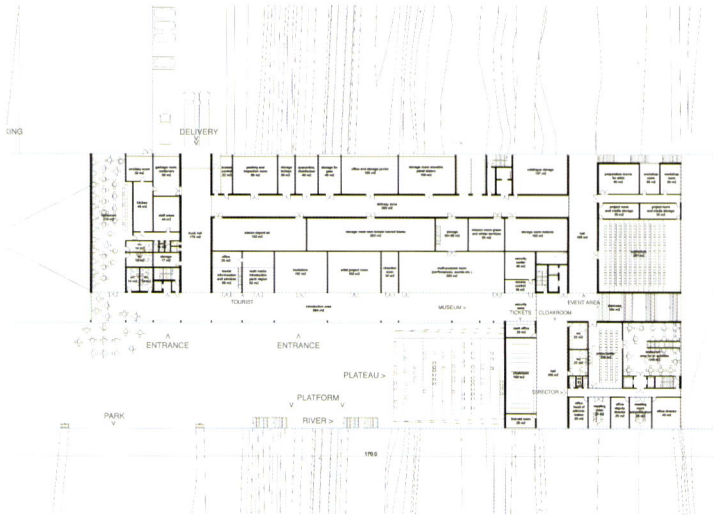

4 LEVEL PLAN

莫斯科的伯纳斯克尼建筑事务所选择把项目建在河里，造成一种层叠效果，甚至要在河堤底部铺设铁轨，建成个车站。

Rooftop perspective 屋顶

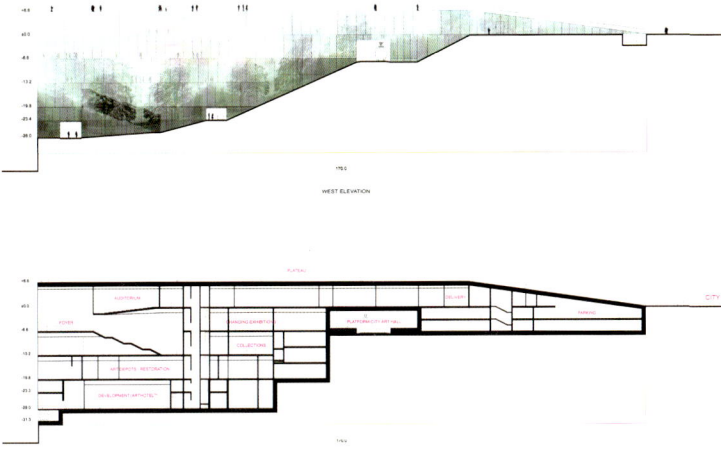

WEST ELEVATION

Section 剖面图

Train station 火车站

瓦雷里奥·奥尔基亚蒂建筑事务所
参赛项目2（并列一等奖）

Valerio Olgiati Project 2
the first prize (2 of 2)

Valerio Olgiati chose to stack the program vertically on the bluff, embellishing the facade with a Frank Lloyd Wright motif.

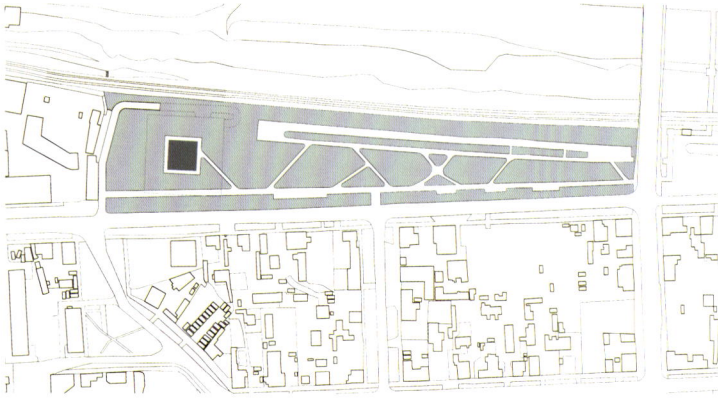

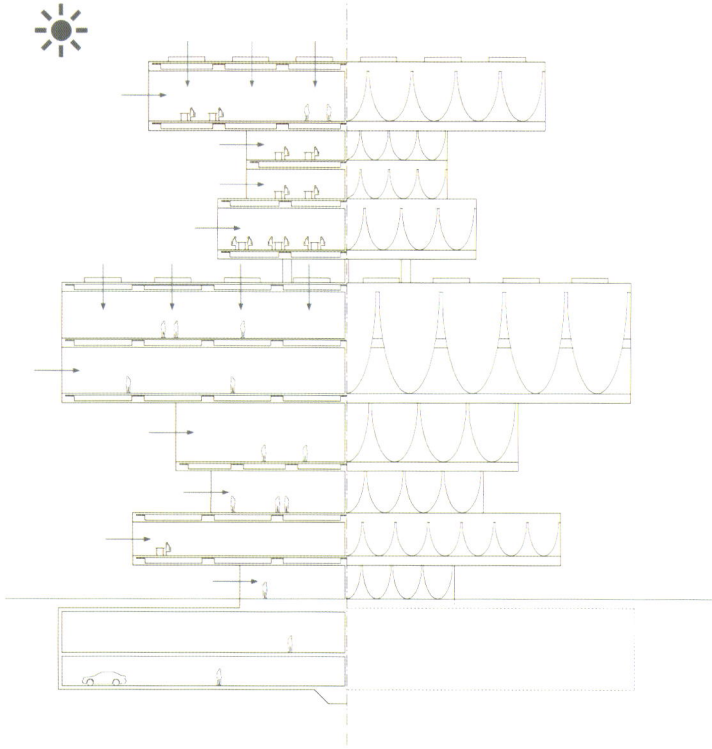

瓦雷里奥·奥尔基亚蒂建筑事务所选择把项目建在垂直悬崖之上，用弗兰克·劳埃德·莱特的基本图案装饰正面。

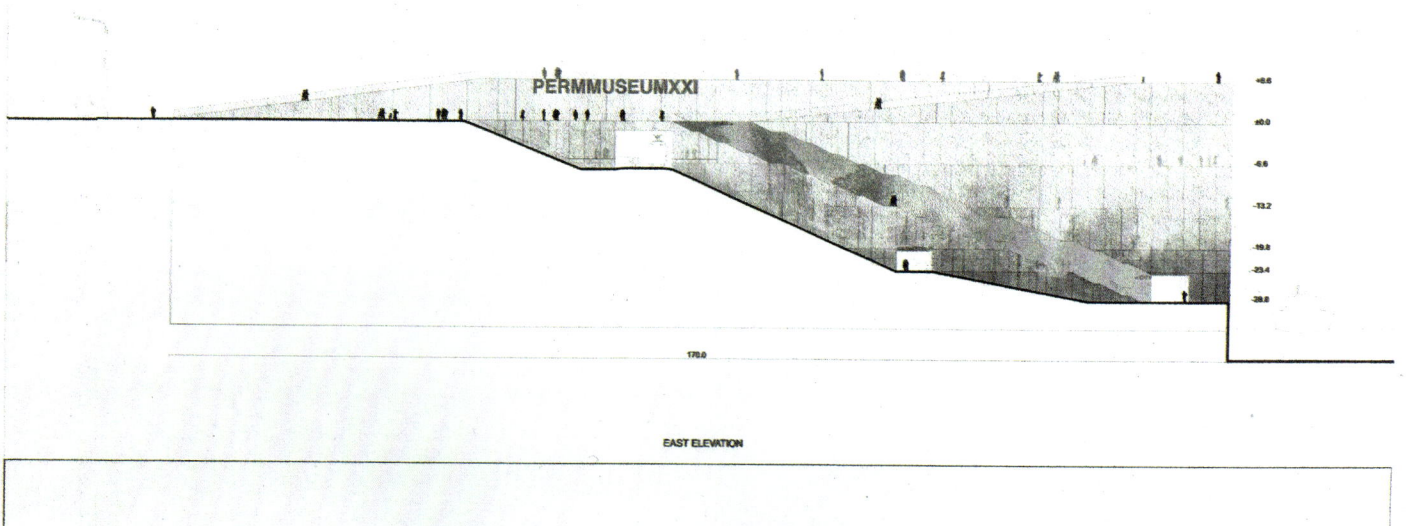

PERMMUSEUMXXI

EAST ELEVATION

扎哈·哈迪德建筑事务所
参赛项目3（三等奖）

Zaha Hadid Architects Project 3
the third prize

Zaha Hadid chose a linear arrangement in an elliptical form, which ran parallel to the river, also along the top of the bluff.

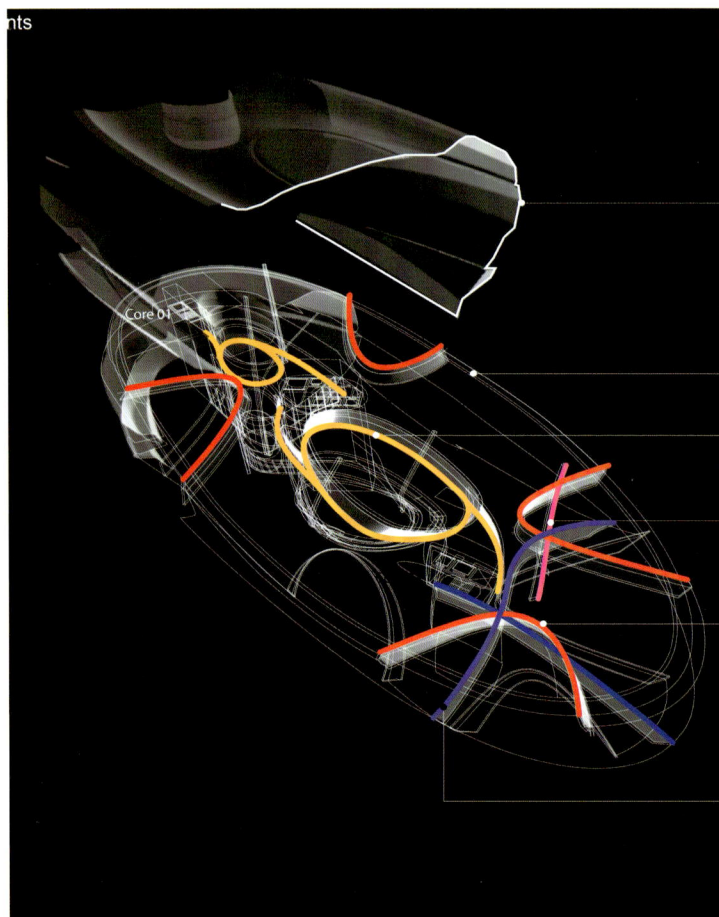

扎哈·哈迪德建筑事务所选择一个线性结构，椭圆形，与河水平行。

Aerial view 鸟瞰图

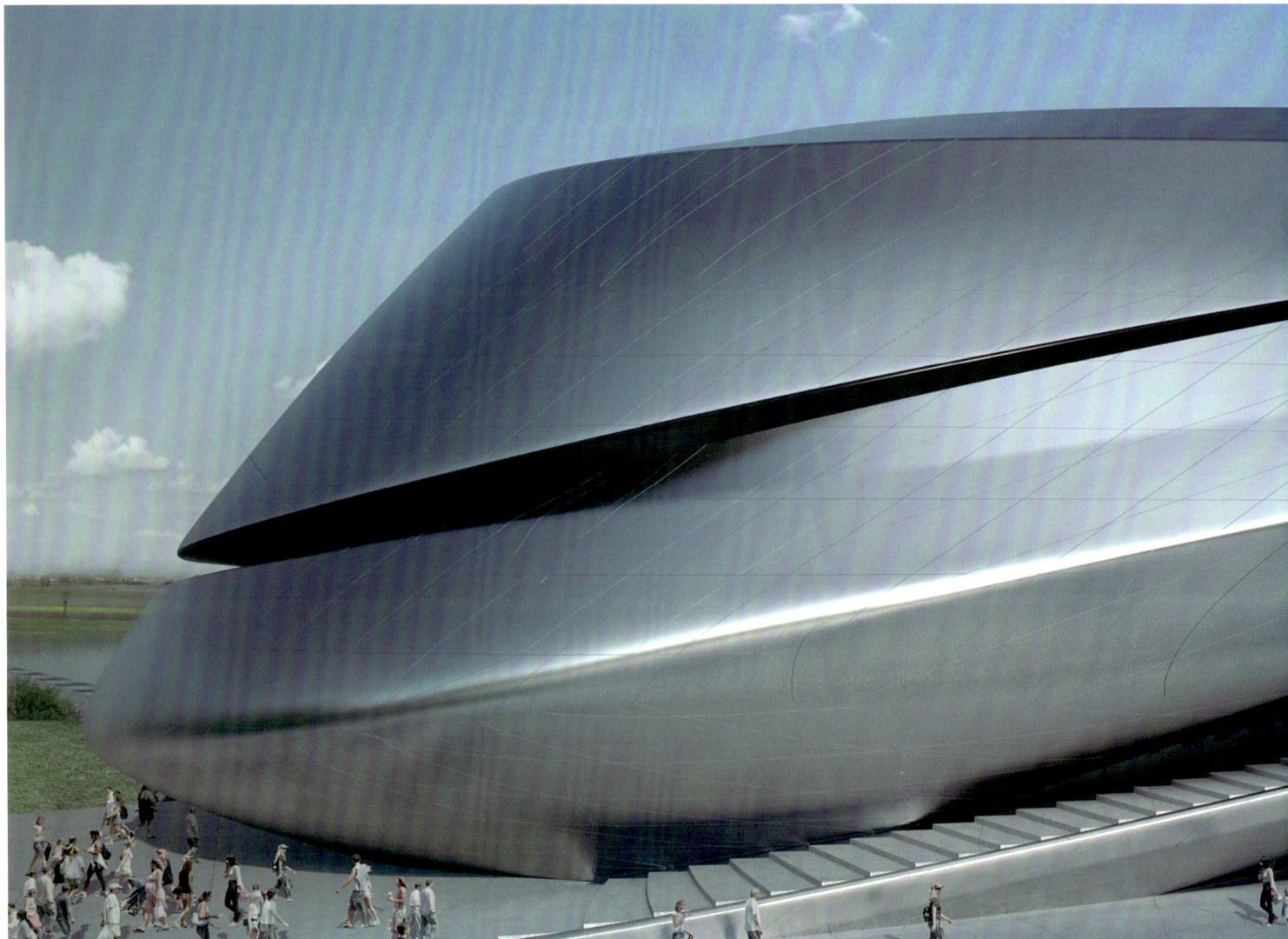
Elevation perspective 立面透视图

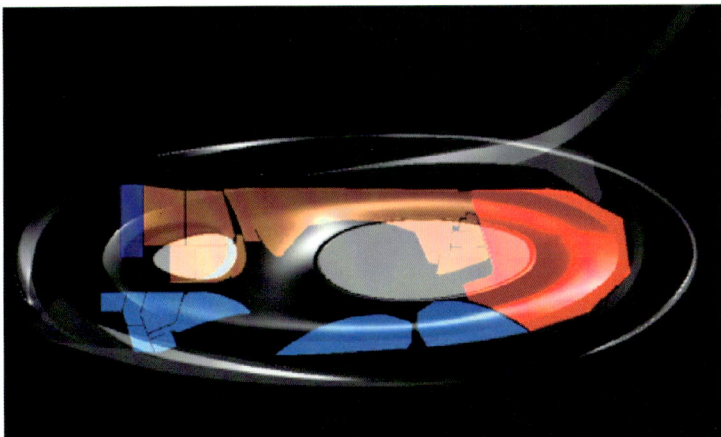

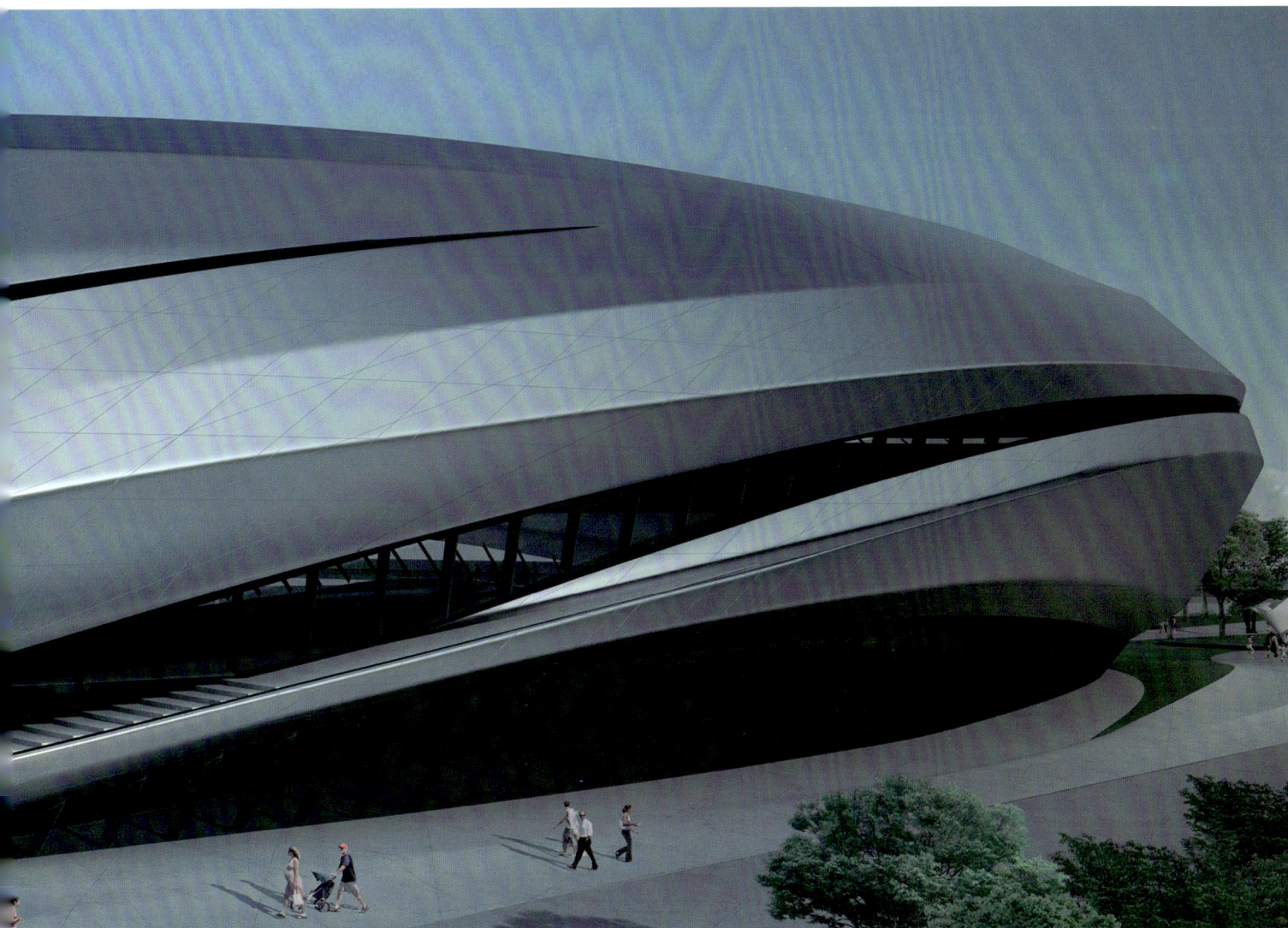

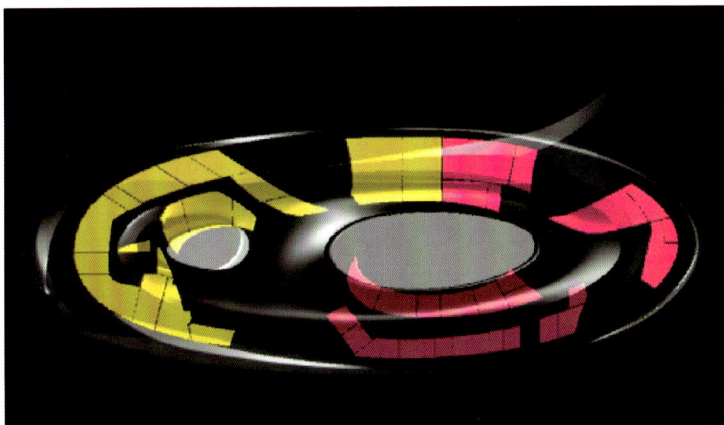

Interior perspective 室内

阿康齐建筑工作室及盖伊诺登森联合建筑事
务所 参赛项目4（特别奖）

Acconci Studio + Guy Nordenson and Associates Project4 special prize

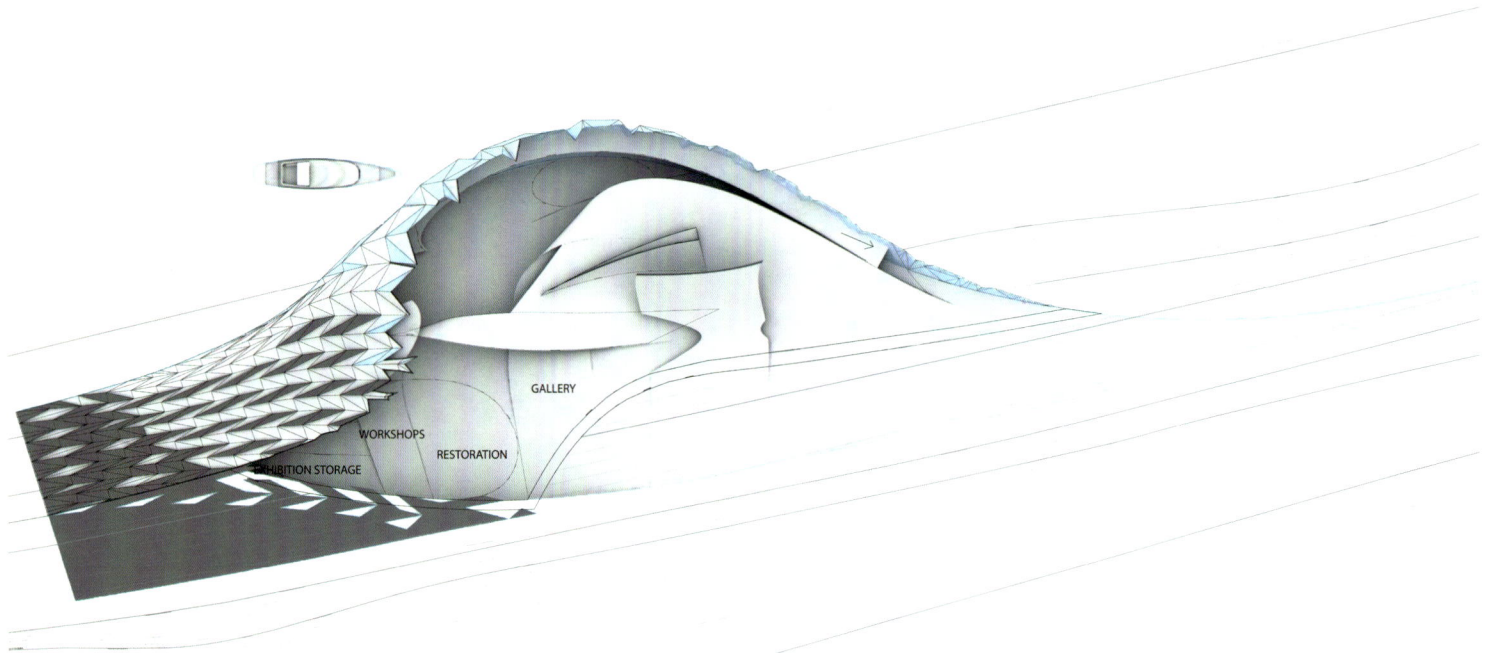

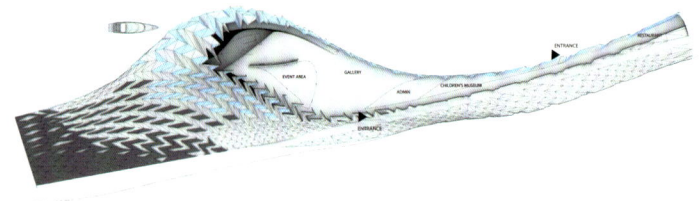

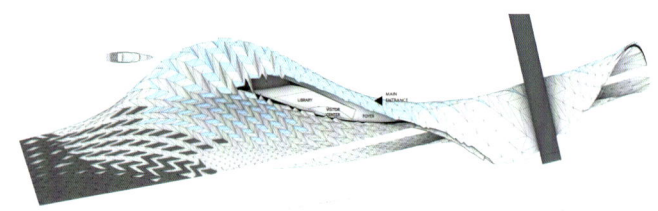

View to entrance 入口处

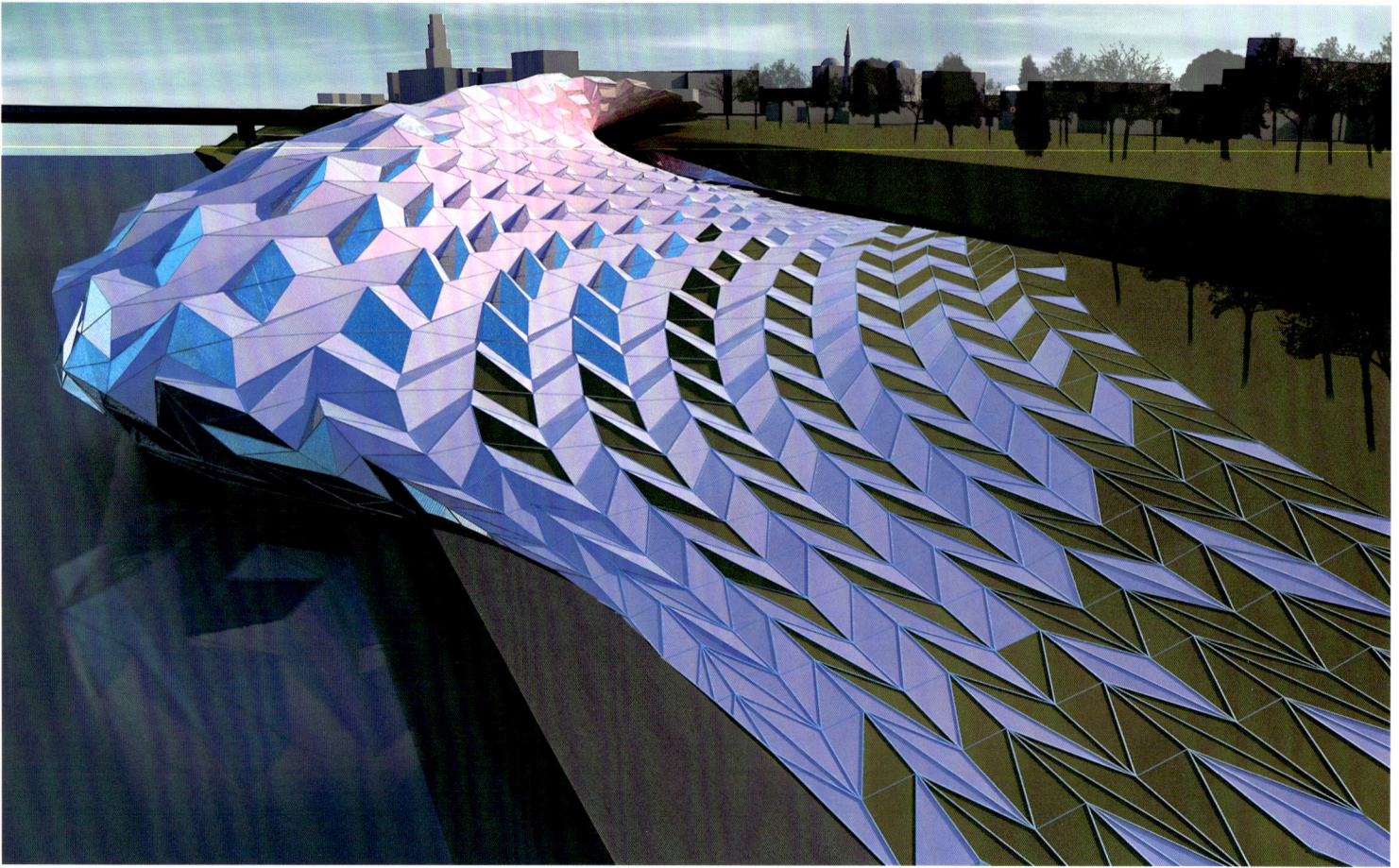

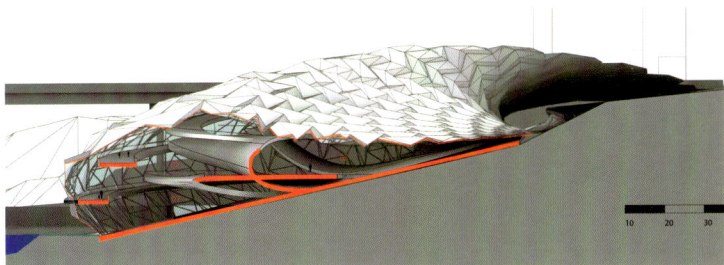

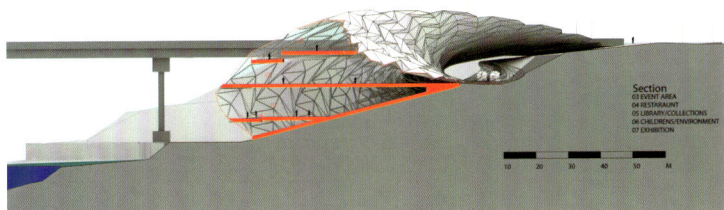

Section 剖面图

亚历山大·布罗德斯基建筑工作室
参赛项目5（特别奖）

Alexander Brodsky Bureau
Project 5 special prize

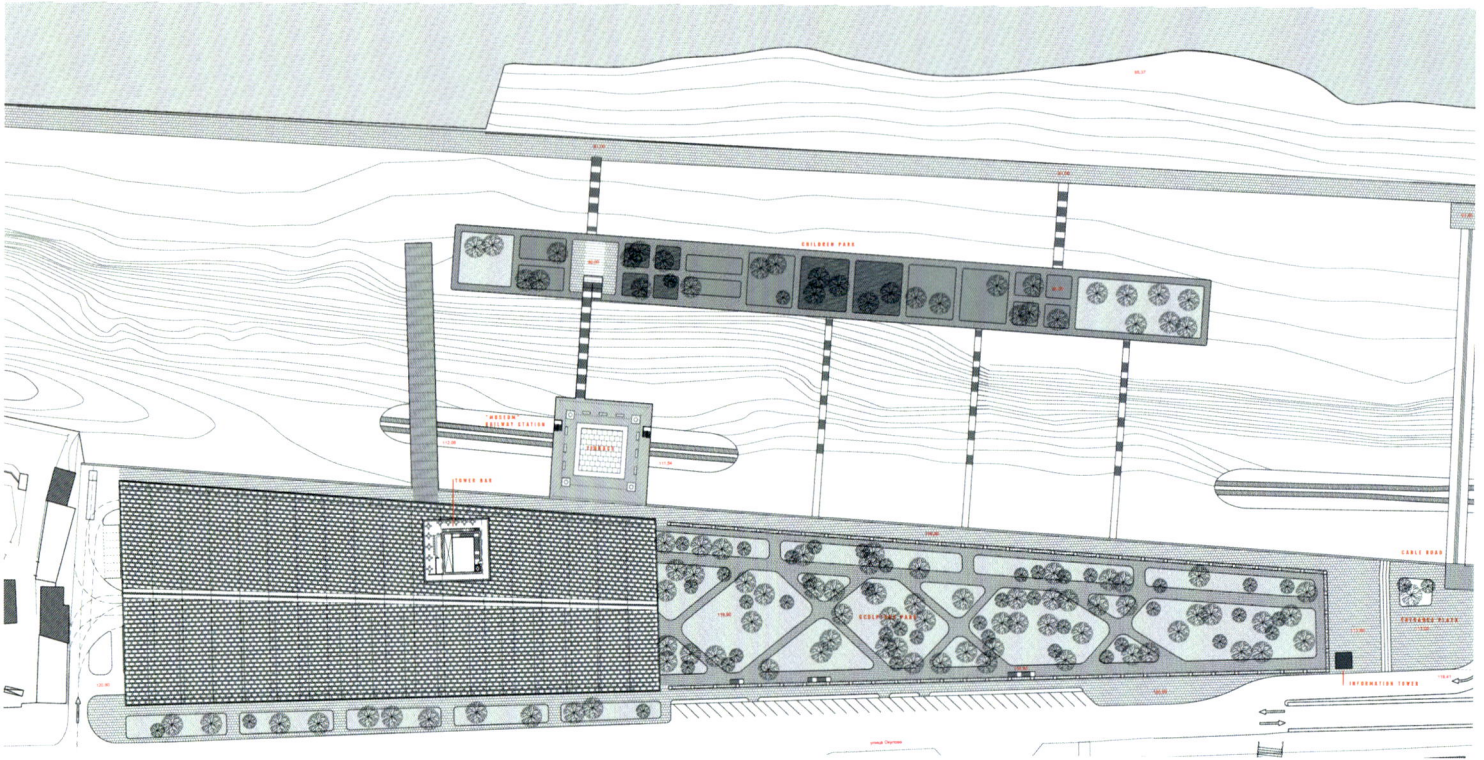

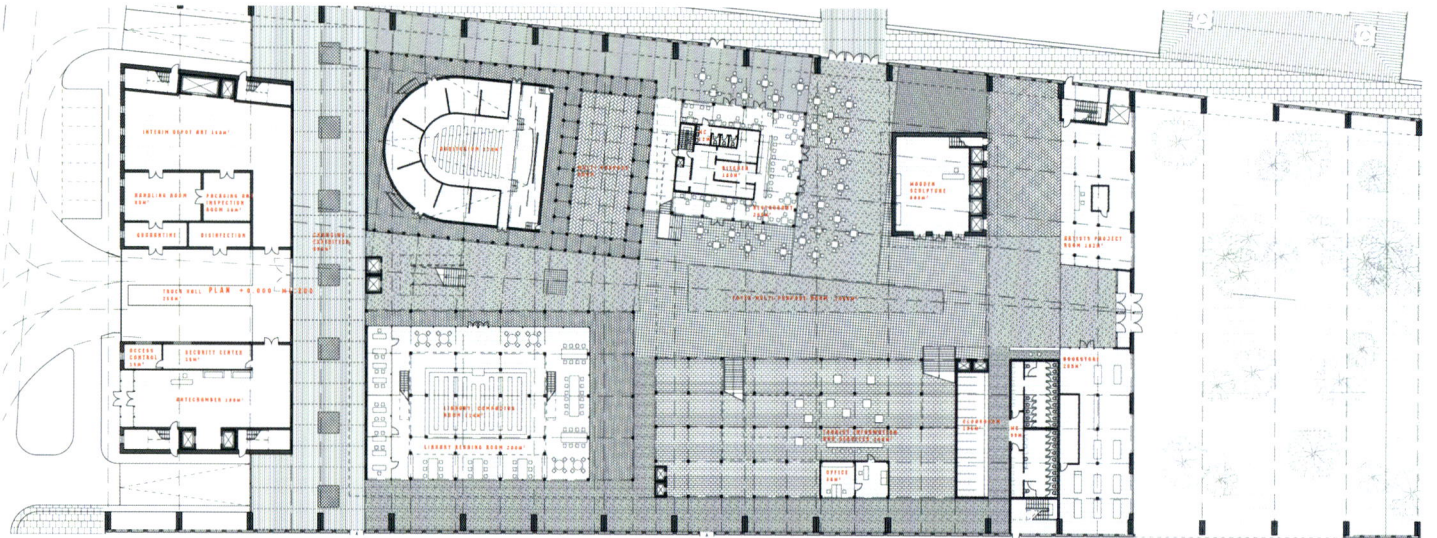

Panorama view 全景图

阿西姆托特建筑事务所
参赛项目6（特别奖）

Asymptote Architecture PLLC
Project 6 special prize

Aerial view to entrance 入口鸟瞰图

Aerial view 鸟瞰图

埃萨·鲁斯基帕 参赛项目7（特别奖） **Esa Ruskeepaa Project 7 special prize**

A – B建筑事务所 参赛项目8（特别奖） **A – B Project 8 special prize**

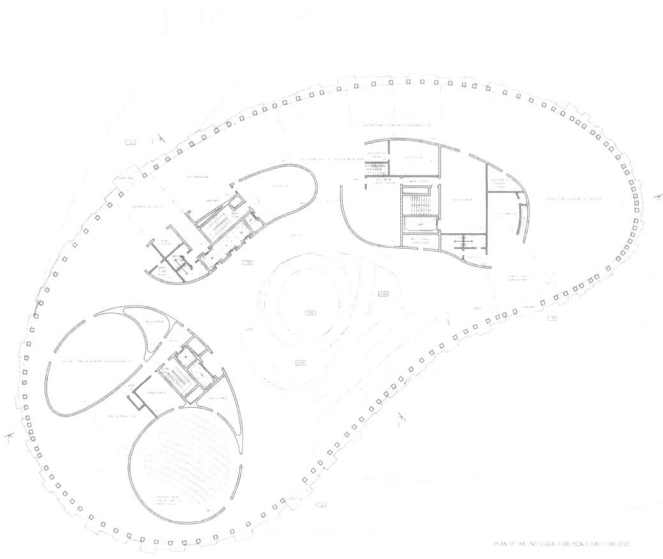

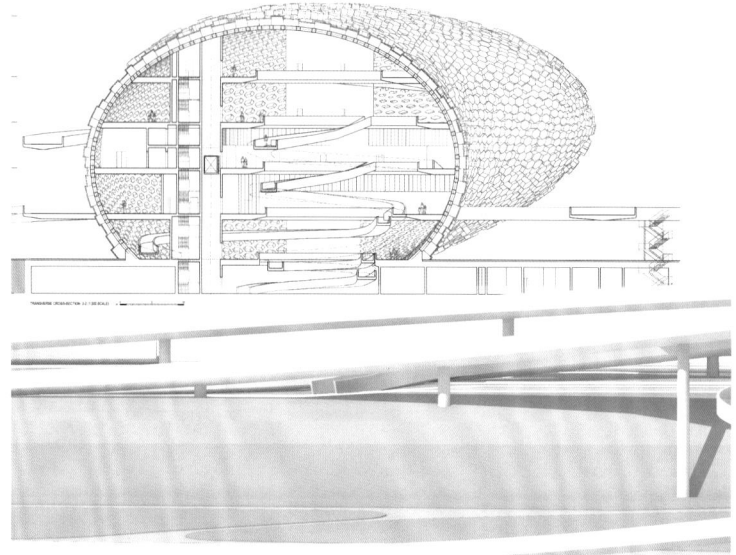

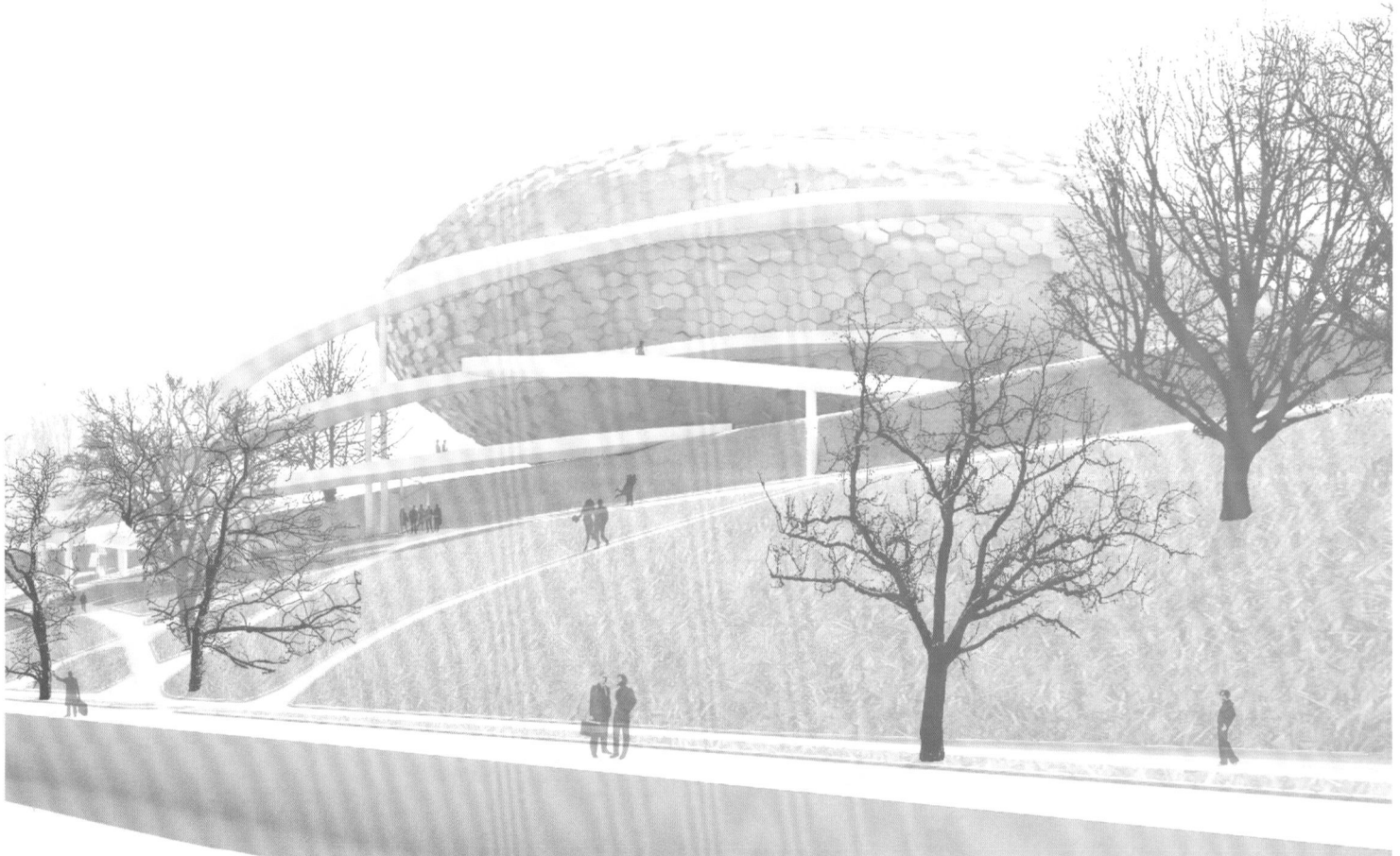

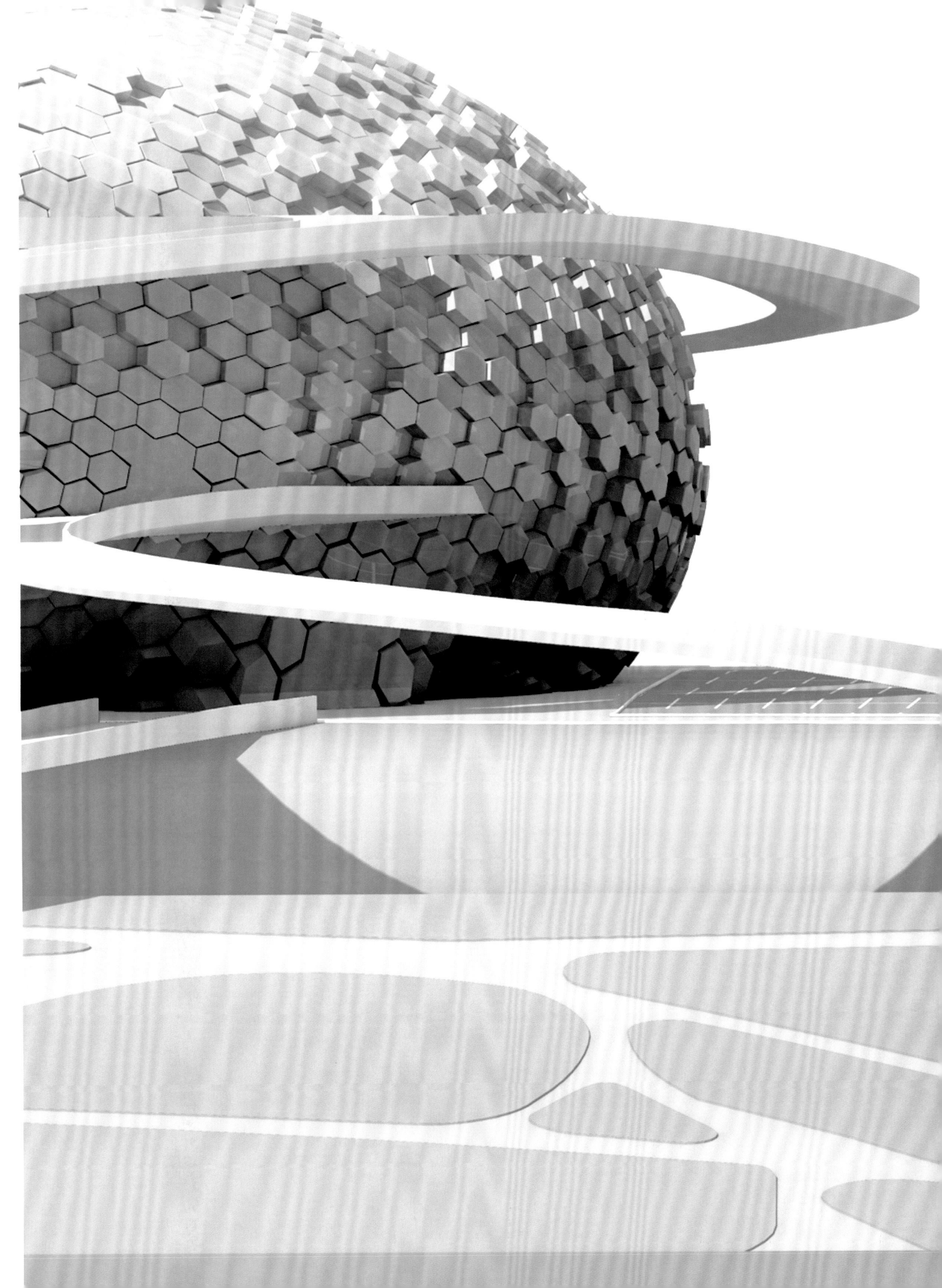

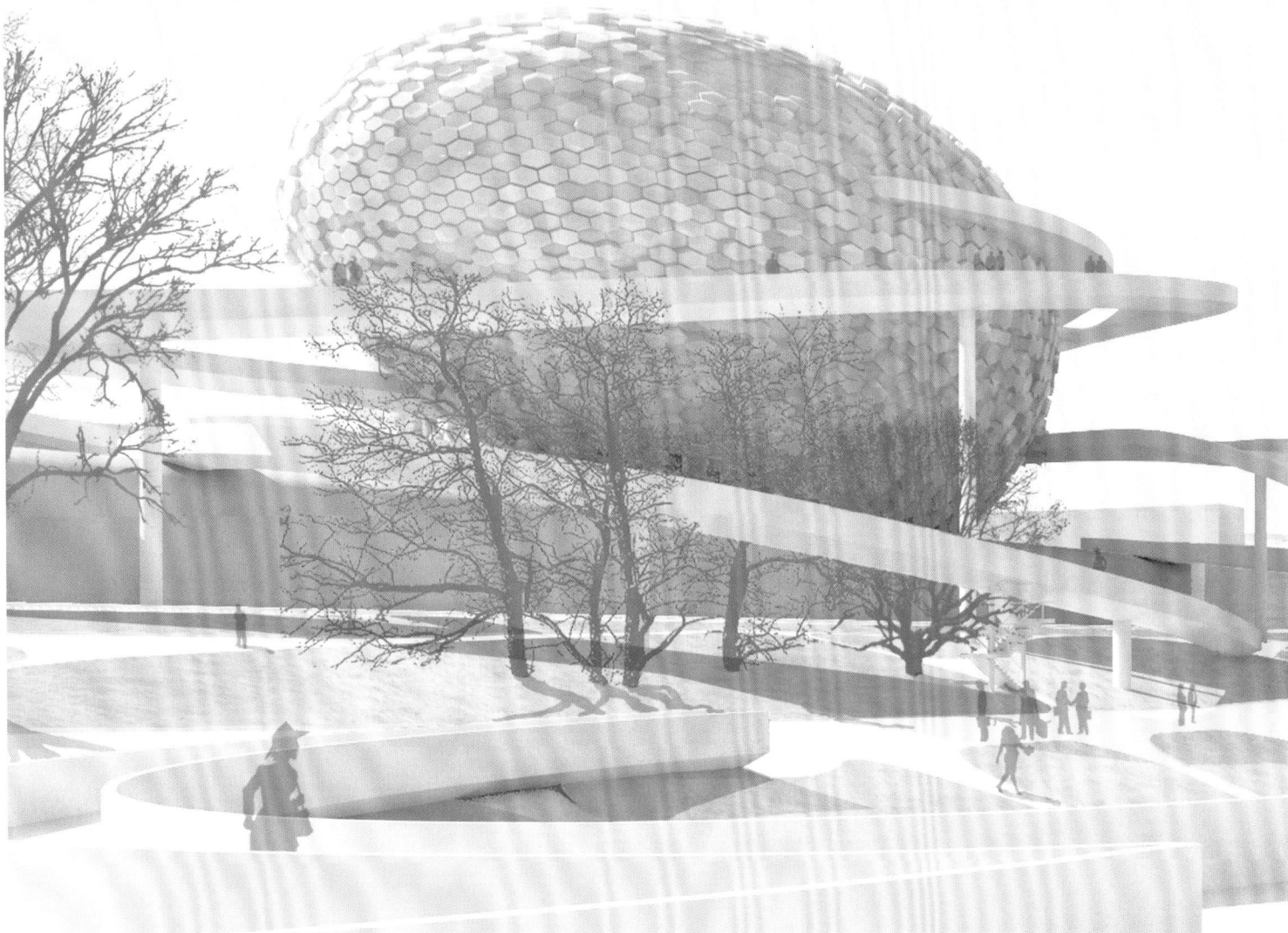

索伦·罗伯特·伦德建筑事务所
参赛项目9（特别奖）

Søren Robert Lund Arkitekter
Project 9 special prize

Aerial view 鸟瞰图

View from railroad 从铁路看博物馆

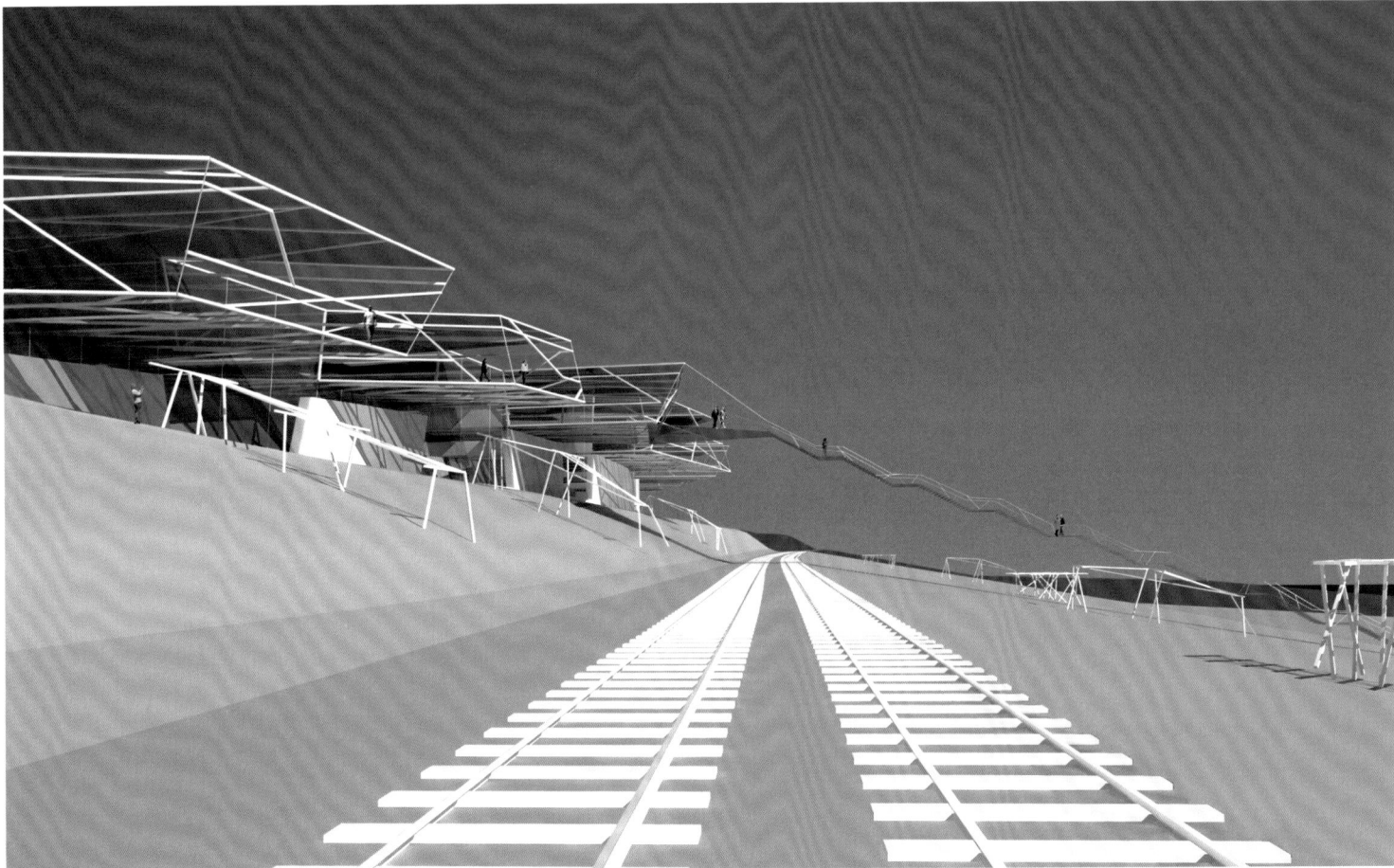

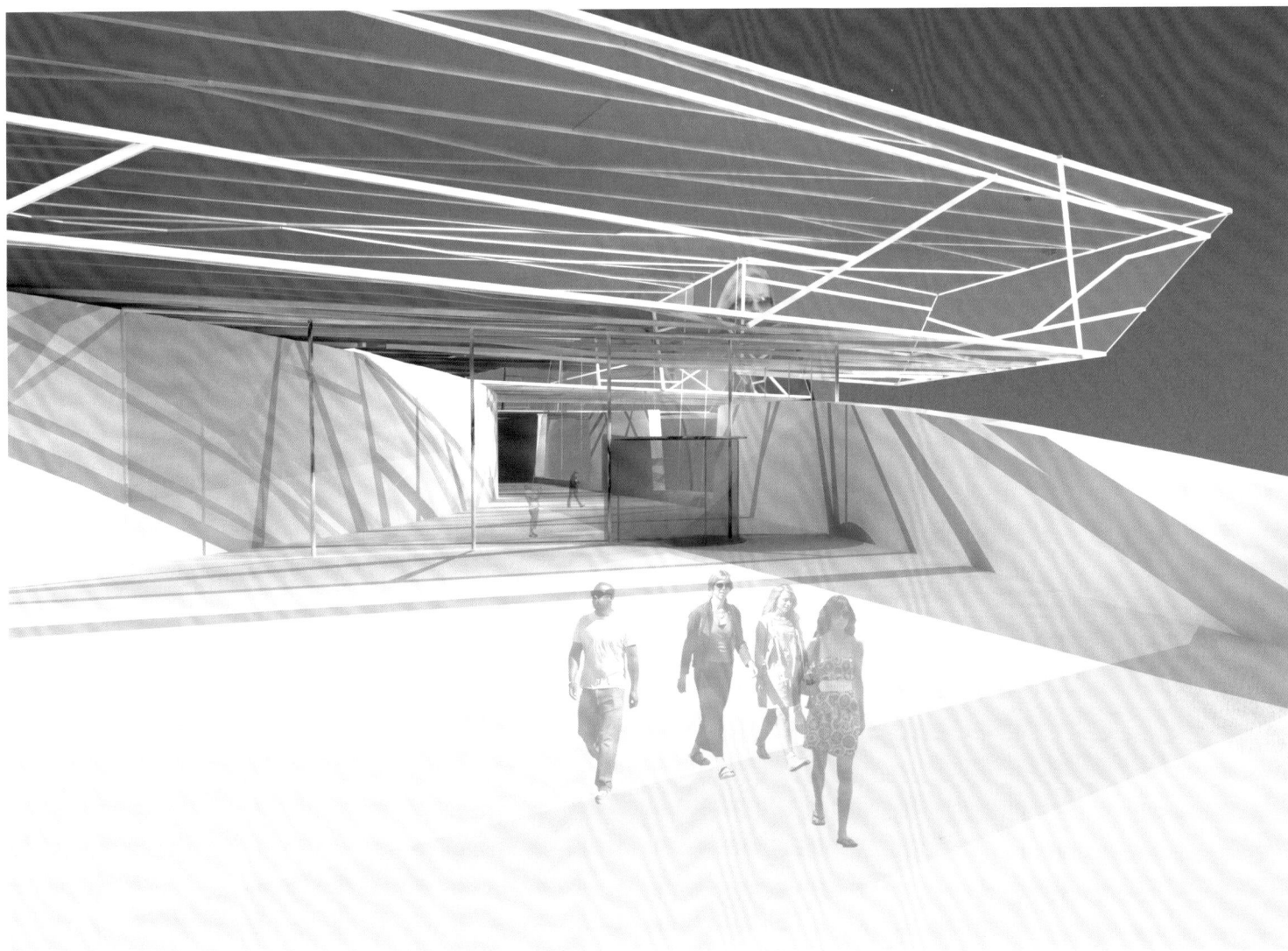

托坦·库泽巴耶夫建筑工作室
参赛项目10（特别奖）

Totan Kuzembaev Architectural Workshop Project 10 special prize

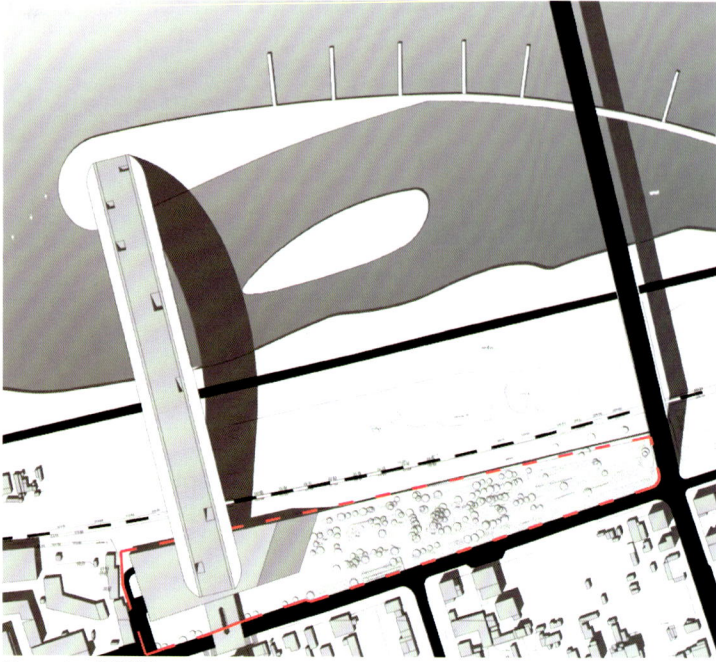

Master plan 总规划图

University of Michigan Museum
Addition Competition

密歇根大学博物馆附楼竞赛

Competition Background

James Stewart, Director of the University of Michigan Museum of Art. Stewart, recently led a competition of sorts for an expansion of his museum, recognized as one of the best university art museums in the country. The University of Michigan Museum of Art is housed in a 42,000 square foot Beaux-Arts building, designed by Donaldson and Meier Architects in 1910. The building was originally conceived as a memorial and is located at the heart of the Ann Arbor campus.

The 55,000 square foot addition will more than double the museum's exhibition space and will also include an auditorium, classrooms, public areas, a museum shop, and conservation and study areas. In addition to serving the campus community, the museum is the only major art collection in Ann Arbor, and the museum hosts a variety of other cultural programs.

密歇根大学艺术博物馆馆长詹姆士·斯图尔特最近牵头举办了一个博物馆扩建项目的竞赛。博物馆坐落在一个3900平方米的美术大楼内，该大楼是由唐纳森和梅尔建筑事务所于1910年设计的。这栋建筑最初被看成是一个纪念馆，位于安·阿巴校园的中心。

拟要修建的5100平方米的附楼，是博物馆的展览空间的一倍之多，还要建一个礼堂、一些教室、公共场所、一家博物馆商店和保护及研究区域。除了为校园社区服务以外，博物馆也是密歇根大学安·阿巴校区主要的艺术收藏馆，用于举办各种各样的其他文化节目。

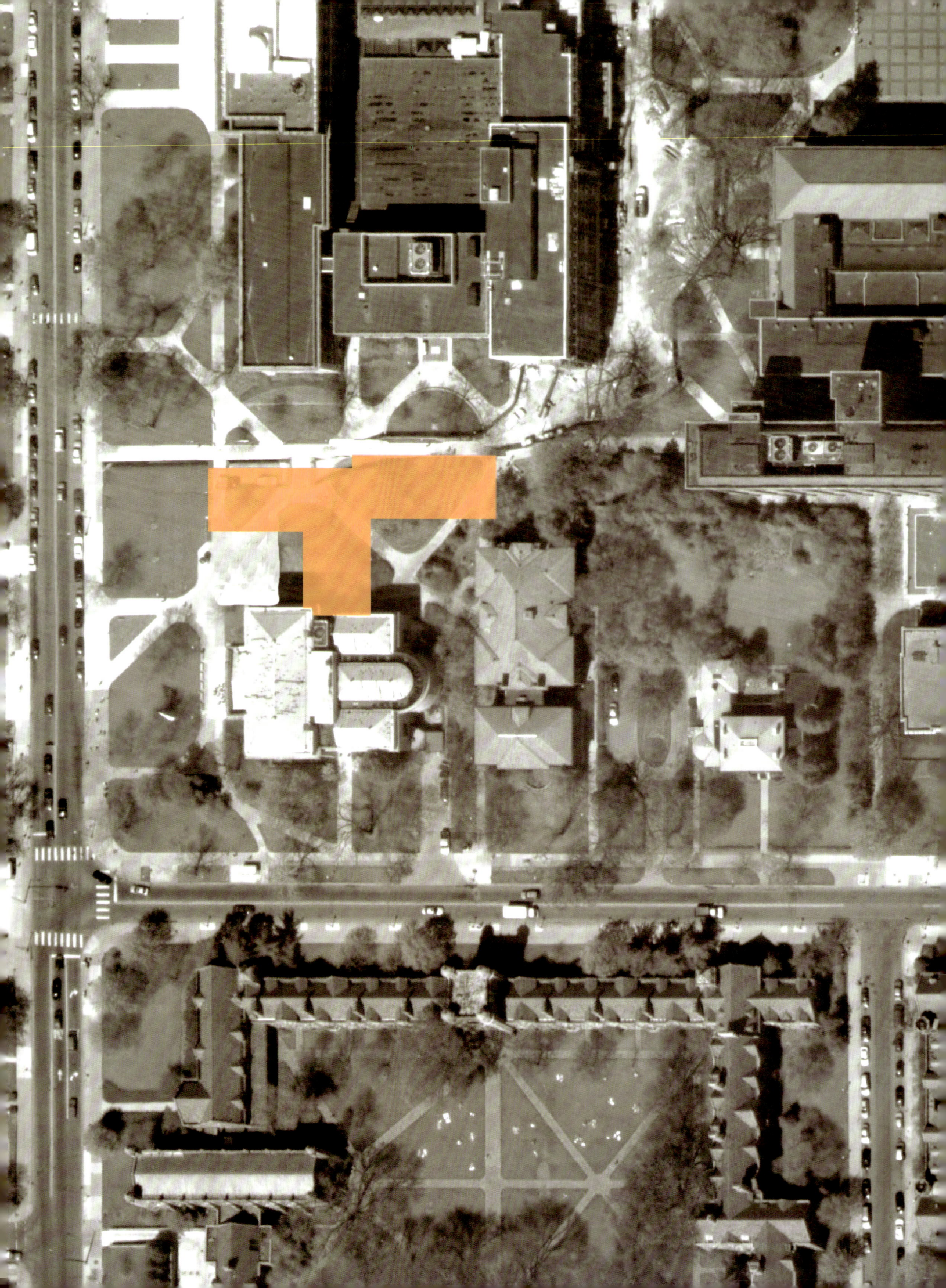

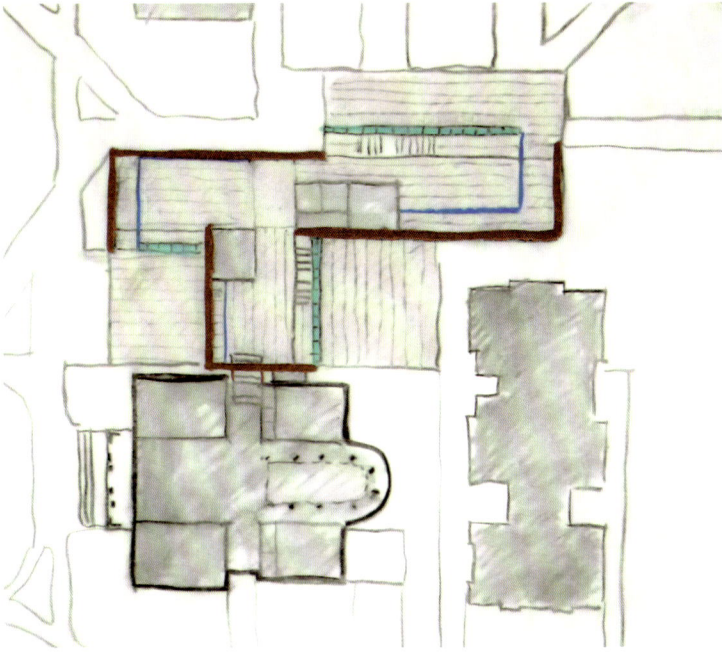

专家点评 Remarks

Building a museum, architecture is as important as, sometimes more so, than what is inside. The entry by Allied Works Architecture not only expressed the respect for this magnificent structure built in 1910, but also realized the harmonious unity of the site and the surrounding.

建筑博物馆，好的建筑有时候要比内部的展示内容更为重要。密歇根大学博物馆附楼竞赛选定的联合经营建筑公司方案，不仅表现了对这个1910年建成的雄伟建筑的尊敬，而且也实现了项目地点与周围环境的和谐。

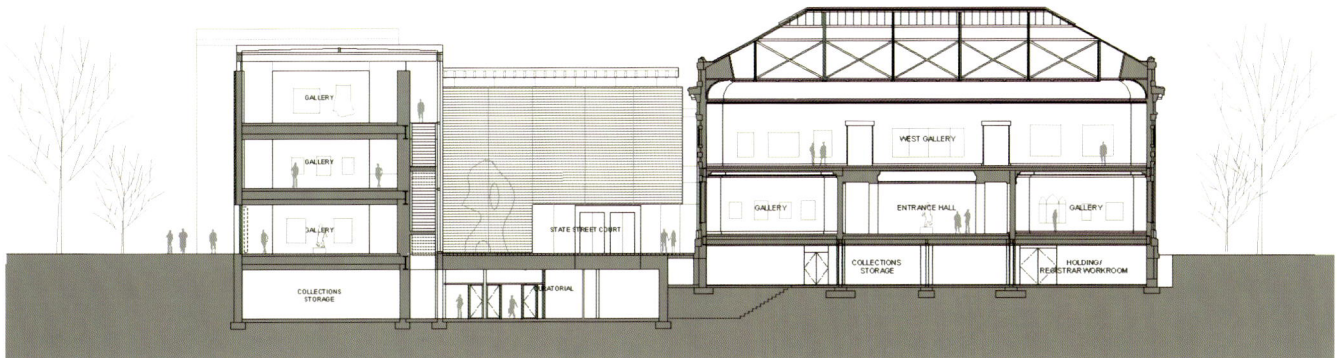

联合经营建筑公司 参赛项目1（一等奖）

Allied Works Architecture
Project 1 the first prize

According to Allied Works's design, students will be able to pass through the T-shaped addition without paying admission, preserving the existing circulation route. The climate in Michigan is extreme enough that students use both indoor and outdoor paths on the campus depending on the season.

Though the materials have yet to be finalized, the team envisions a façade treatment that includes vertical bands of glass alternating with heavier columns of stone and steel, in deference to the monumental facades of the 1910 building. While the building will be distinctly modern, this rhythmic façade treatment echoes the classicism of Memorial Hall, while allowing for more transparency and a sense of activity for the viewer outside. The T-shaped addition will also create new outdoor spaces, which could be used for a variety of curatorial purposes, thus creating an integrated relationship between indoor and out, institution and campus.

Ground floor plan 一层平面图

联合经营建筑公司的设计方案，学生无须付费就可以通过T型附楼，从而维护了校园内现有的流通路线。密歇根的气候变化大，学生可以根据季节使用校园内的室内和室外路径。

虽然材料有待敲定，联合经营建筑公司的设计团队想出了一种包括垂直玻璃条和石头钢架结构交替使用的建筑物正面处理方式，表现了对这个1910年建成的雄伟建筑的尊敬。虽然该建筑将会特别现代，但是这种带有韵律的建筑正面在提供建筑通透感和动感的同时，也会与纪念馆的古典风格互相呼应。这个T型附楼建筑也将创造出新的户外空间，可用于各种策展用途，因此，在室内和室外之间，在机构与校园之间，创造出一种综合一体的关系。

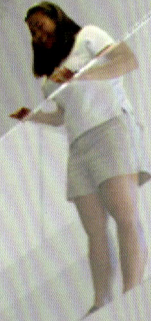

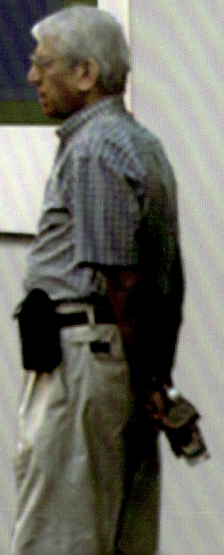

Lobby perspective 门厅透视图

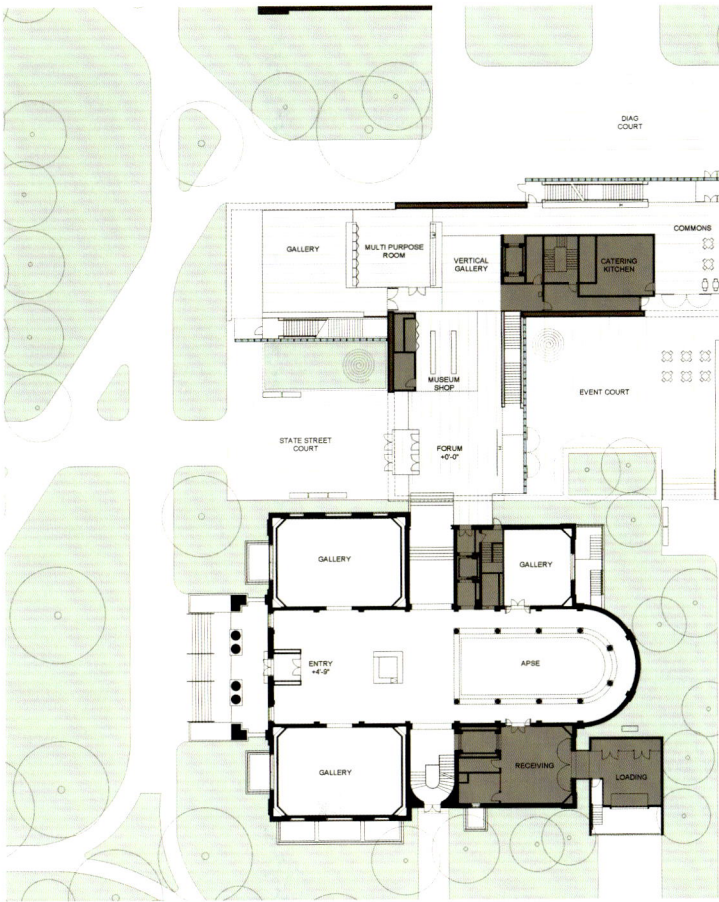

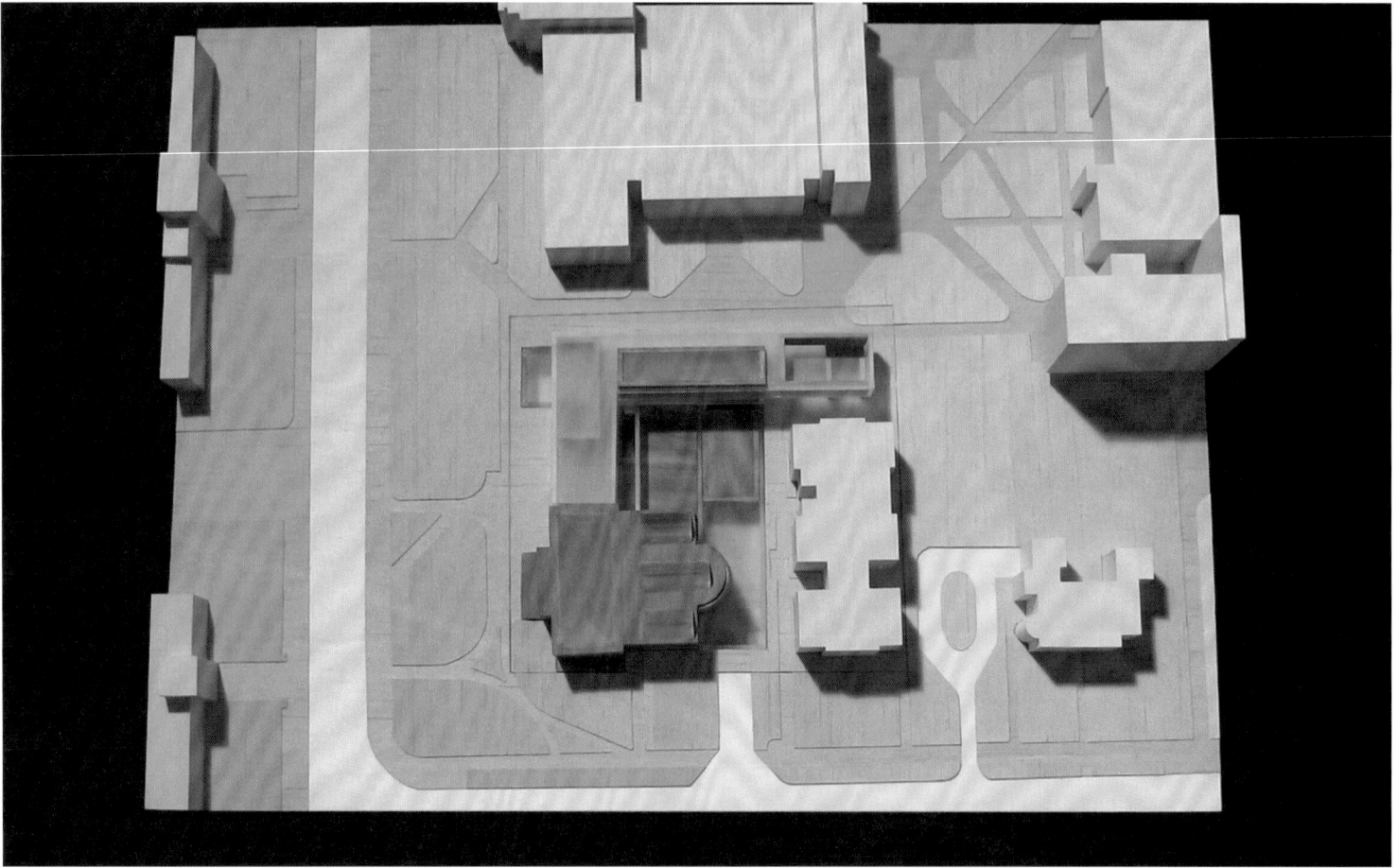
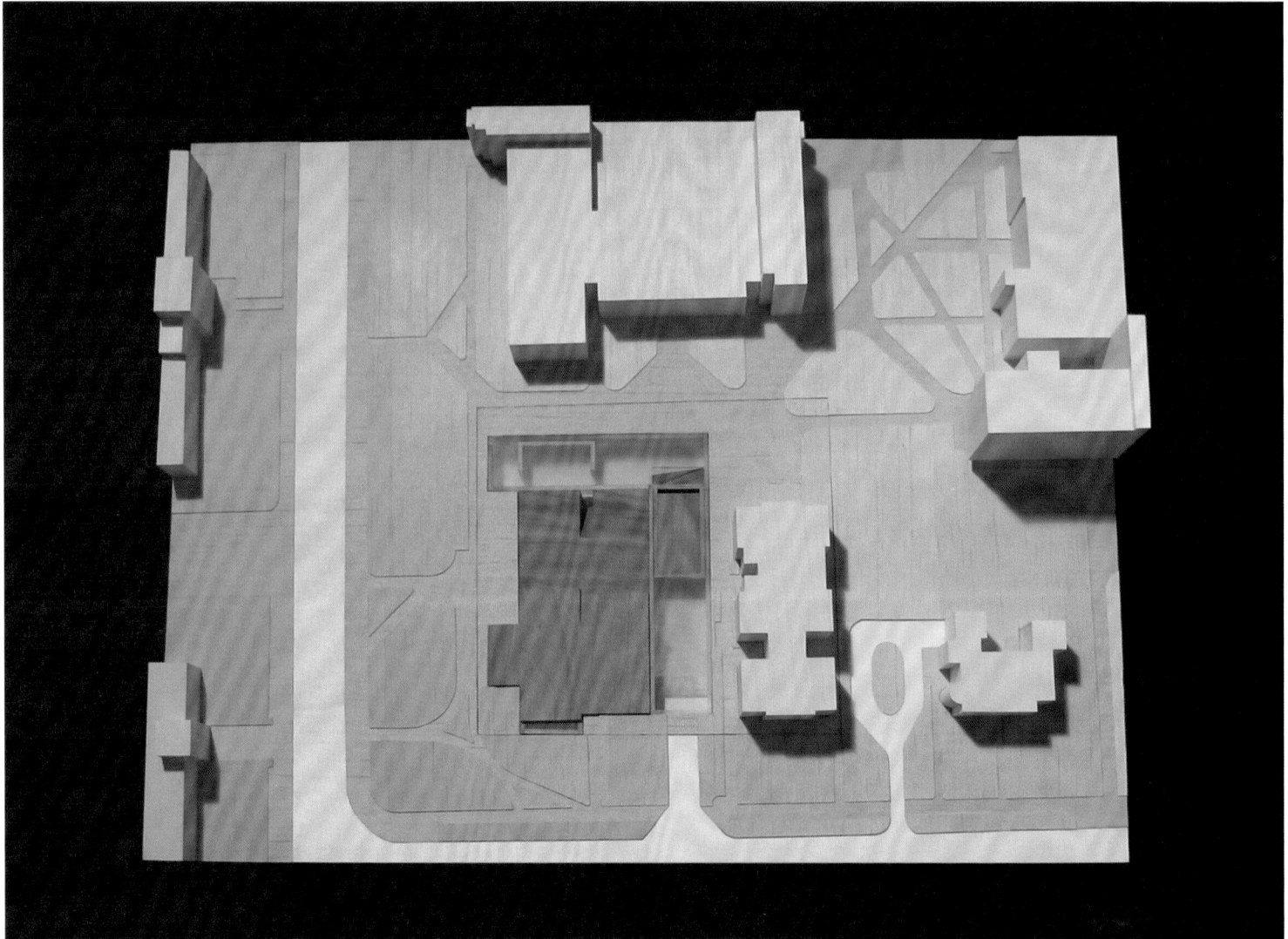

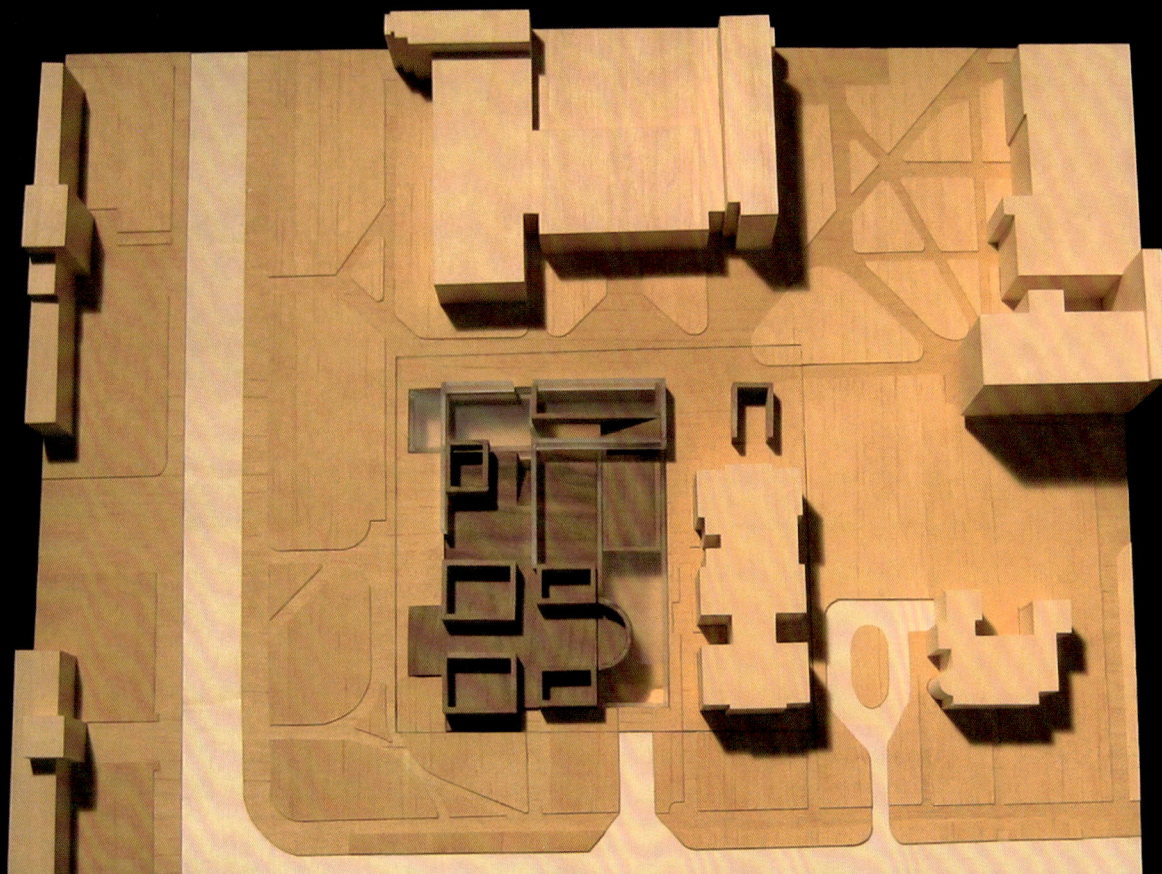

波尔夏克及合作伙伴建筑事务所
参赛项目2（入围奖）

The Polshek Partnership
Project 2 finalist

The Polshek Partnership proposed a box within a box, one transparent and animated, the other opaque. The outer box would be rendered in glass and steel, allowing for students to pass through the building, without entering the galleries themselves. Public areas would be placed within this outer box, helping to activate the space both for those inside and for those viewing the building from the outside. "We were filling a void on the campus that in many ways didn't want to be filled," says Todd Schliemann, design partner at Polshek, "So the idea of perimeter circulation is a way to both dematerialize the building and to make it come alive." The highly transparent facades would also allow for large banners or electronic screens to be visible to announce exhibitions or public events. The inner box, which Polshek designed in a cool stone, to impart a greater degree of gravitas, would be quieter, and more contemplative.

This reflects Stewart's belief that museums should be more than spectacles, that they should create a sense of sanctuary, and a place for unhurried experiences.

Model shot with exterior skin 带有外表面的模型镜头

波尔夏克及合作伙伴建筑事务所的方案是一个盒中盒的概念。一个透明，栩栩如生；另一个晦暗。外盒是玻璃和钢的材质，学生不需要进入画廊便可在建筑物中通过。公共区域将被设置在这个外盒内部，有助于使空间充满动感。外围流通的设计概念既弱化了建筑形态，又使它变得活灵活现。高度透明的建筑物正面使大标语或电子屏幕清晰可见，用来就展览或公共活动进行宣传。波尔夏克用凉爽石头设计内盒结构，以便更大程度地赋予其一种庄严感，建成之后会显得更加安静，引人沉思。

该方案反映了斯图尔特的信念，即博物馆不光要建得壮观，还应该创造出一种圣洁感，让人们获得一种悠闲的经验。

Façade perspective with Guernica image 带有格尼卡图案的正面透视图

Concourse with ramp 带斜坡的大厅

韦斯·曼弗瑞迪建筑事务所
参赛项目3（入围奖）

Weiss/Manfredi Architects Project 3 finalist

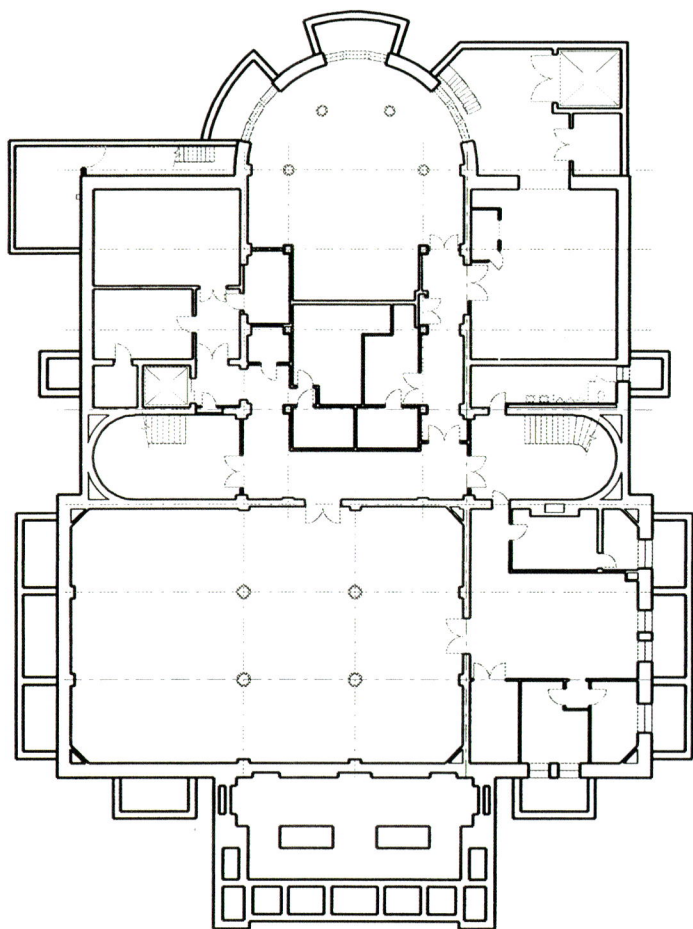

Weiss/Manfredi, as with much of their work, proposed a building that combined dramatic form with a careful consideration of site and landscape. They proposed a curvilinear building that would acknowledge the traditional paths of pedestrian circulation (over which the new addition is being built). Students would also be allowed to pass through the Weiss/Manfredi building, which placed the public areas on the ground floor, and the galleries on the second floor.

The primary intent was to create a series of deflected, curved spaces that acknowledge the way the building fit in the context of the campus. In some way the galleries would respond in a pedagogic way. Their smooth, curved form would stand in deliberate contrast to the more rectilinear space of the Beaux-Arts building.

韦斯·曼弗瑞迪建筑事务所的方案提出了要把建筑的戏剧化形式与对项目地点及景观的仔细考虑结合起来。他们建议采用曲线型建筑，认可行人步道的传统路径。学生们也可以穿过韦斯·曼弗瑞迪大楼通行，该大楼建在公共场所一楼，而画廊位于二楼。

设计的主要意图是要建立一系列的认可建筑与校园背景契合的曲线空间。画廊在某种程度上会起到教育的作用。光滑的曲线形式将与美术大楼更多的多边形空间形成鲜明对比。

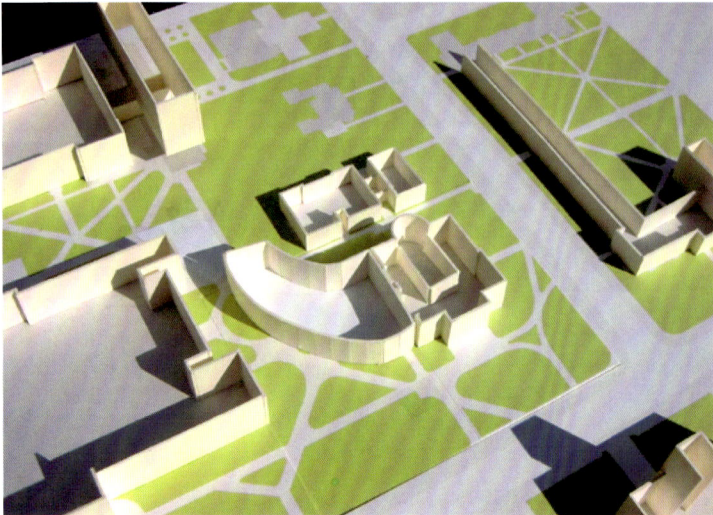

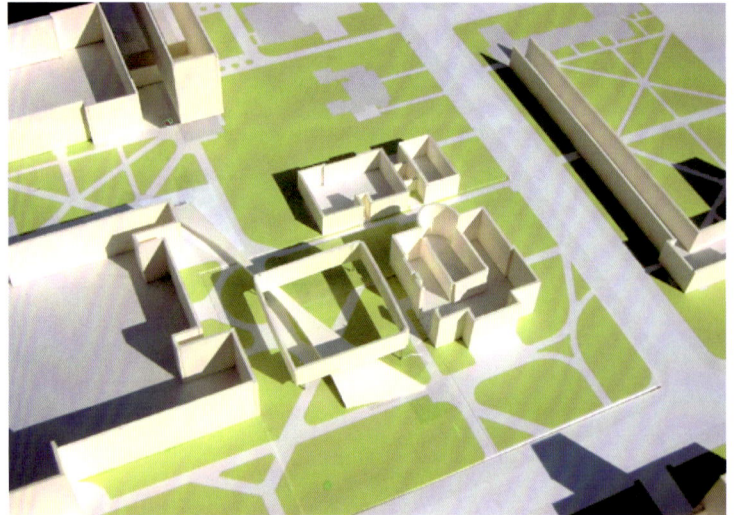

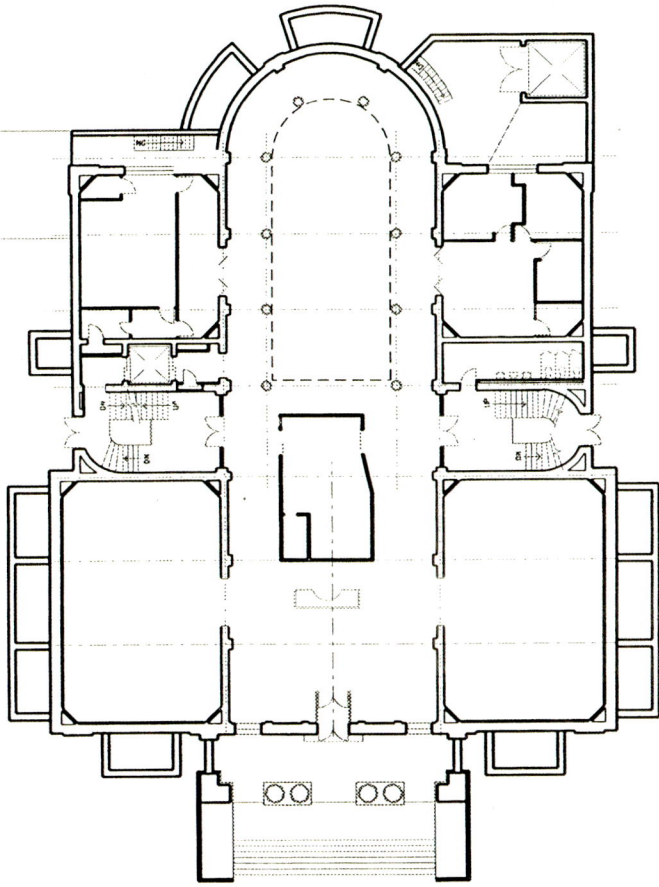

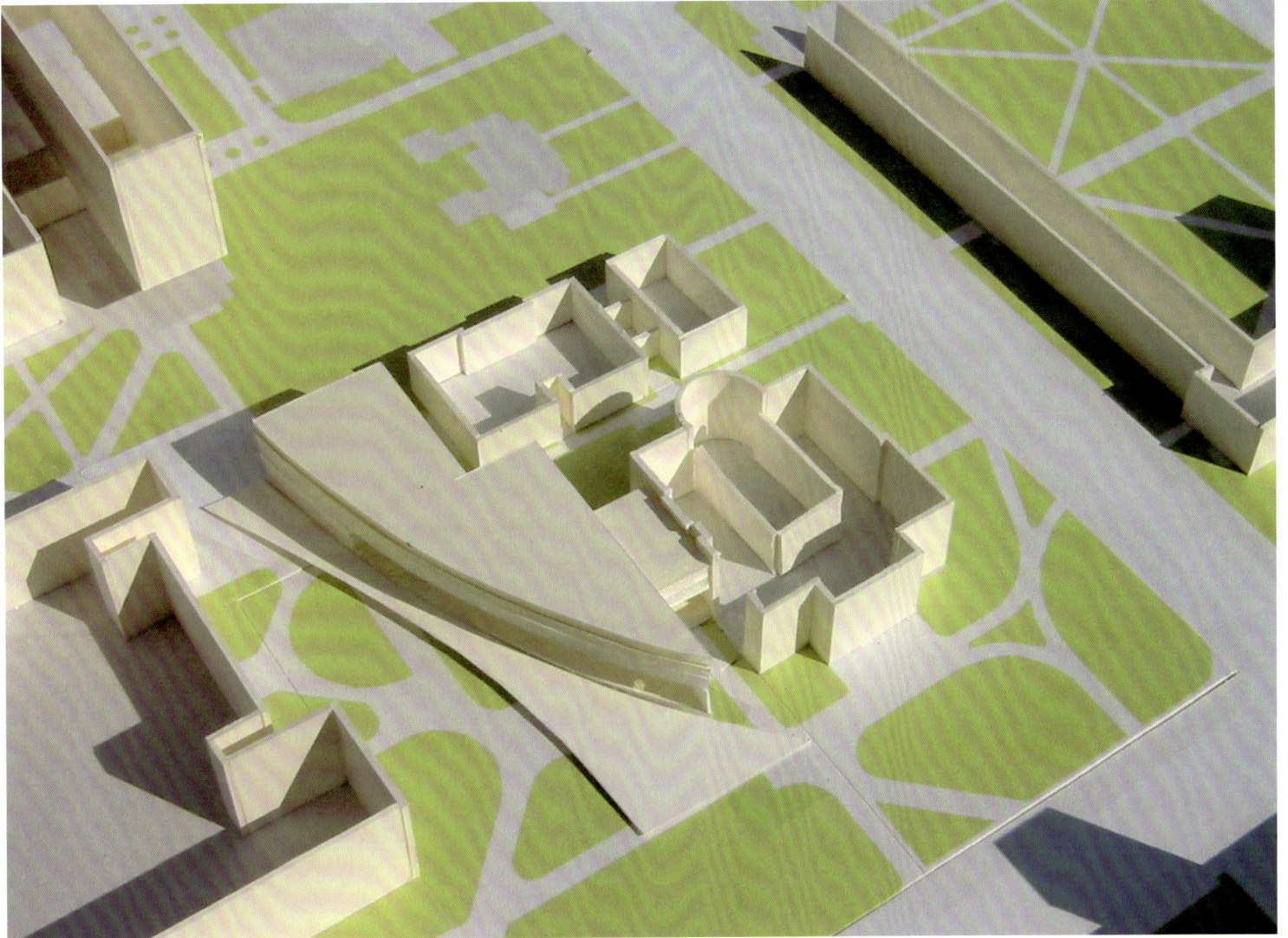

Whatcom Museum of History & Art and Whatcom Children's Museum Competition

霍特科姆艺术历史博物馆和儿童博物馆竞赛

竞赛背景 Competition Background

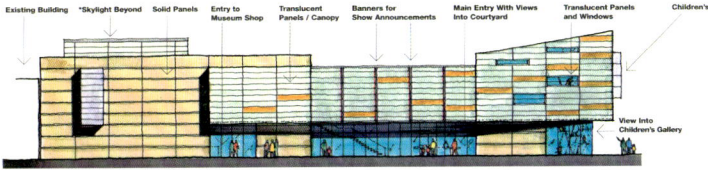

Ever since Gehry's Bilbao Guggenheim, communities of all sizes have prioritized high design as they seek to build new cultural and civic facilities. Bellingham, Washington is no exception. When the Bellingham-Whatcom Public Facilities District and the City of Bellingham set out to build a 35,000-square-foot, $8 million building as the new home for the Whatcom Museum of History & Art and the Whatcom Children's Museum, they had an icon in mind to serve as the linchpin of a downtown cultural district. Concurrently, they decided that a design competition would be the best vehicle to achieve their goals.

自从毕尔巴鄂古根海姆博物馆项目取得成功之后，大大小小的社区在兴建新的文化市政建筑时，都开始重视艺术设计性了。华盛顿州的柏林汉姆市也不例外。柏林汉姆－霍特科姆郡公共设施区和柏林汉姆市政府计划出资800万美元，建造占地3250平方米的霍特科姆艺术历史博物馆和儿童博物馆。他们致力要把这座博物馆打造成城区文化枢纽，形成一个城市地标。而举办一次设计竞赛是达到目的的最佳途径。

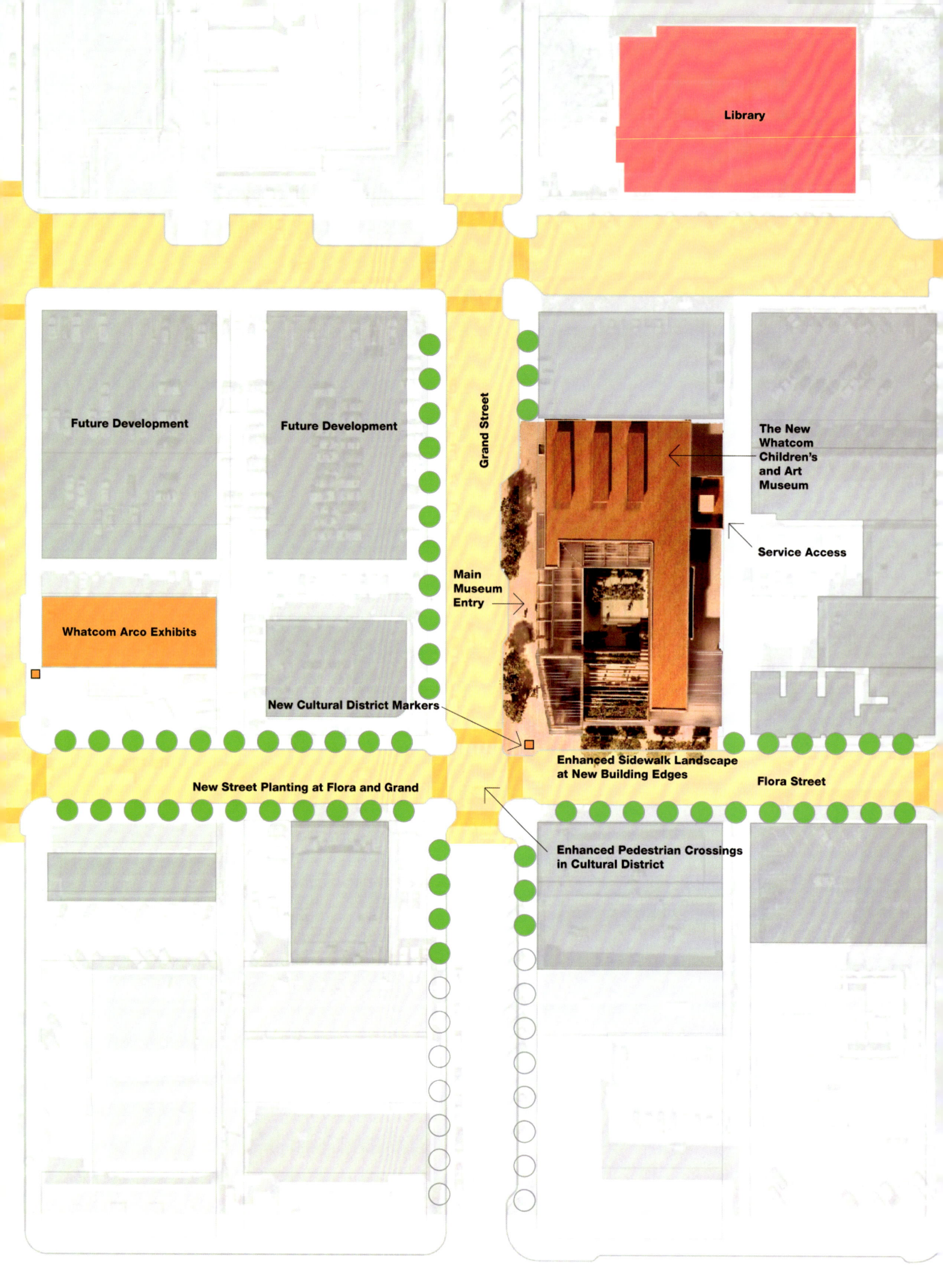

Library

Future Development

Future Development

Grand Street

The New Whatcom Children's and Art Museum

Service Access

Whatcom Arco Exhibits

Main Museum Entry

New Cultural District Markers

Enhanced Sidewalk Landscape at New Building Edges

Flora Street

New Street Planting at Flora and Grand

Enhanced Pedestrian Crossings in Cultural District

专家点评 **Remarks**

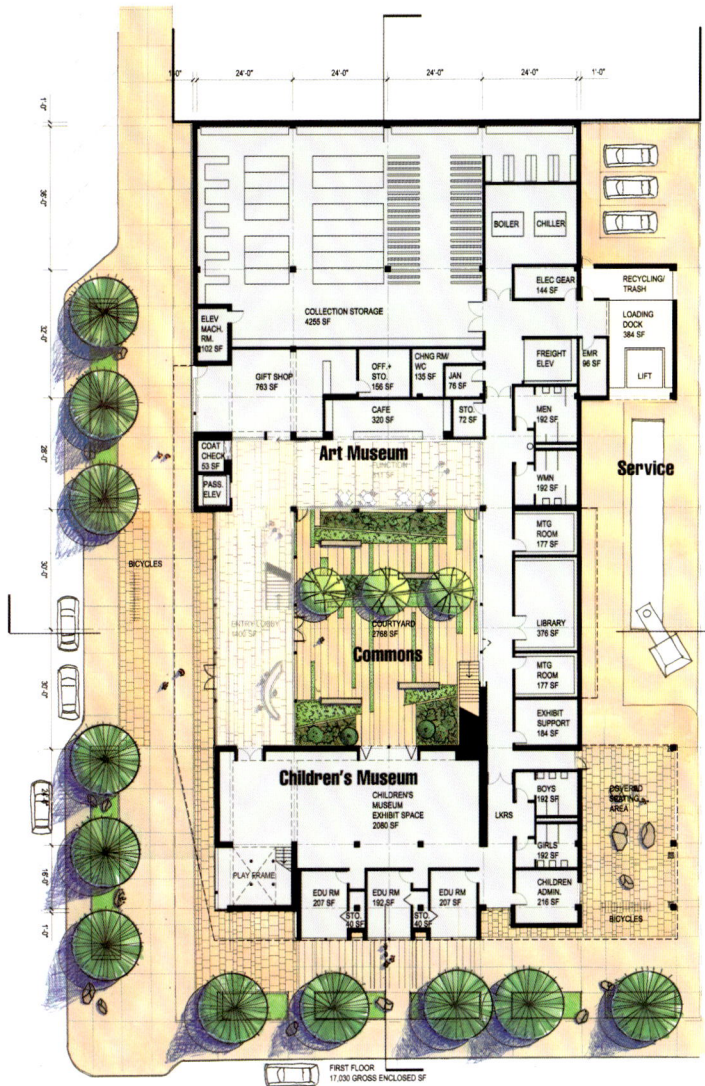

Level 1 Plan

According to the jury Alison Williams, all the three designs are excellent, while Olson's scheme is more refreshing and more suitable for this community of long history.

"One thing we liked about the Olson scheme was that it was not a fait accompli," she said. "It had a clear, big, well-developed idea, but it wasn't trying to solve everything: its simplicity made it very appealing. It was not monumental, it had an attitude about daylight, and there was a simplicity that left open opportunities so that the client could really engage (throughout) the evolution of the design to the final result." Williams also praised the design team's careful orientation as it related to the precious nature of light in this region and the design's "basic quality" as very appealing to the jury.

评委艾莉森·威廉姆斯称，三个团队的设计都很优秀，但是奥尔森的设计更让人耳目一新，更适合这个具有悠久历史的社区。

她说："我们欣赏奥尔森的设计的一个重要原因就是：它不是一个既成事实。这一设计有清晰、宏大、明确的理念，但是并不急于解决所有问题。简单让它更具吸引力。奥尔森的设计不追求不朽，而是留有余地。简单到让客户可以有更多的机会参与到设计之中。"威廉姆斯还称赞设计团队对地区本质的准确定位和设计的"基本质量"吸引了评审委员会的注意。

剑桥七建筑事务所/唐纳利建筑事务所
参赛项目1（入围奖）

Cambridge Seven Associates, Inc./Donnally Architects
Project 1 finalist

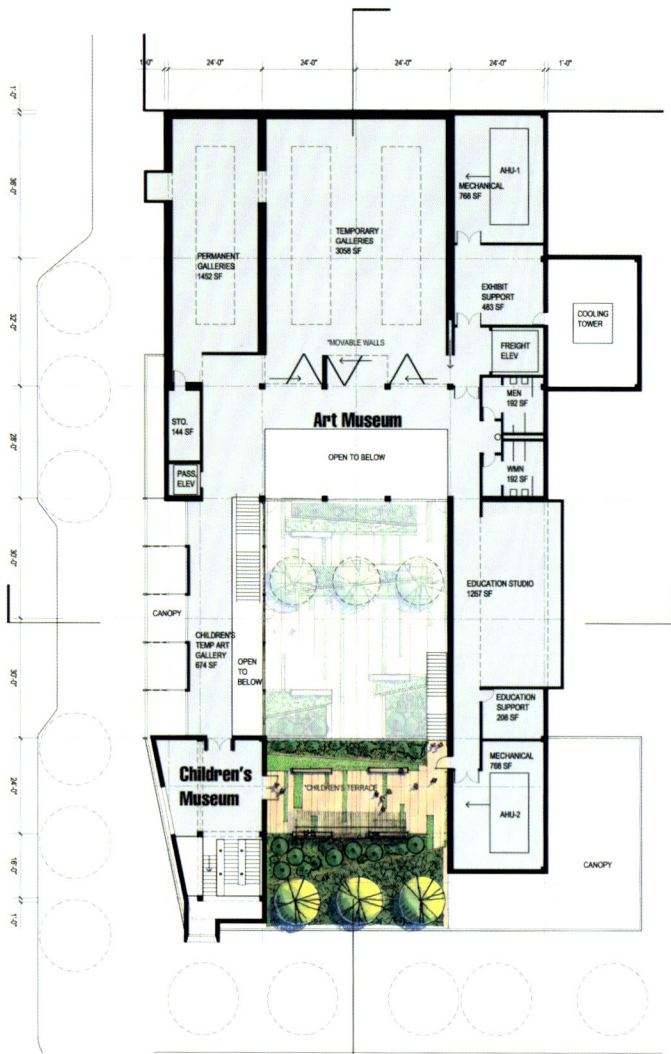

Floor plan 平面图

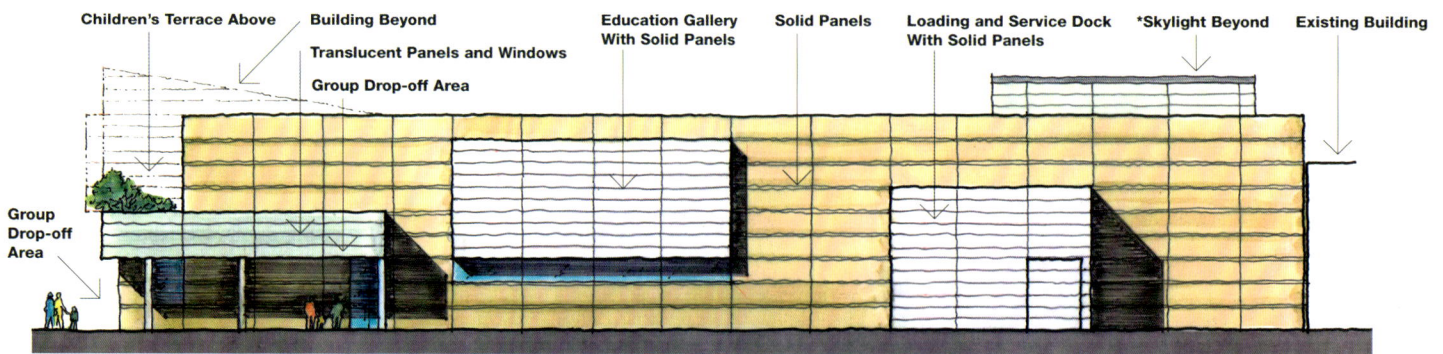

East elevation 东立面图

Birdeye view of model 鸟瞰图模型

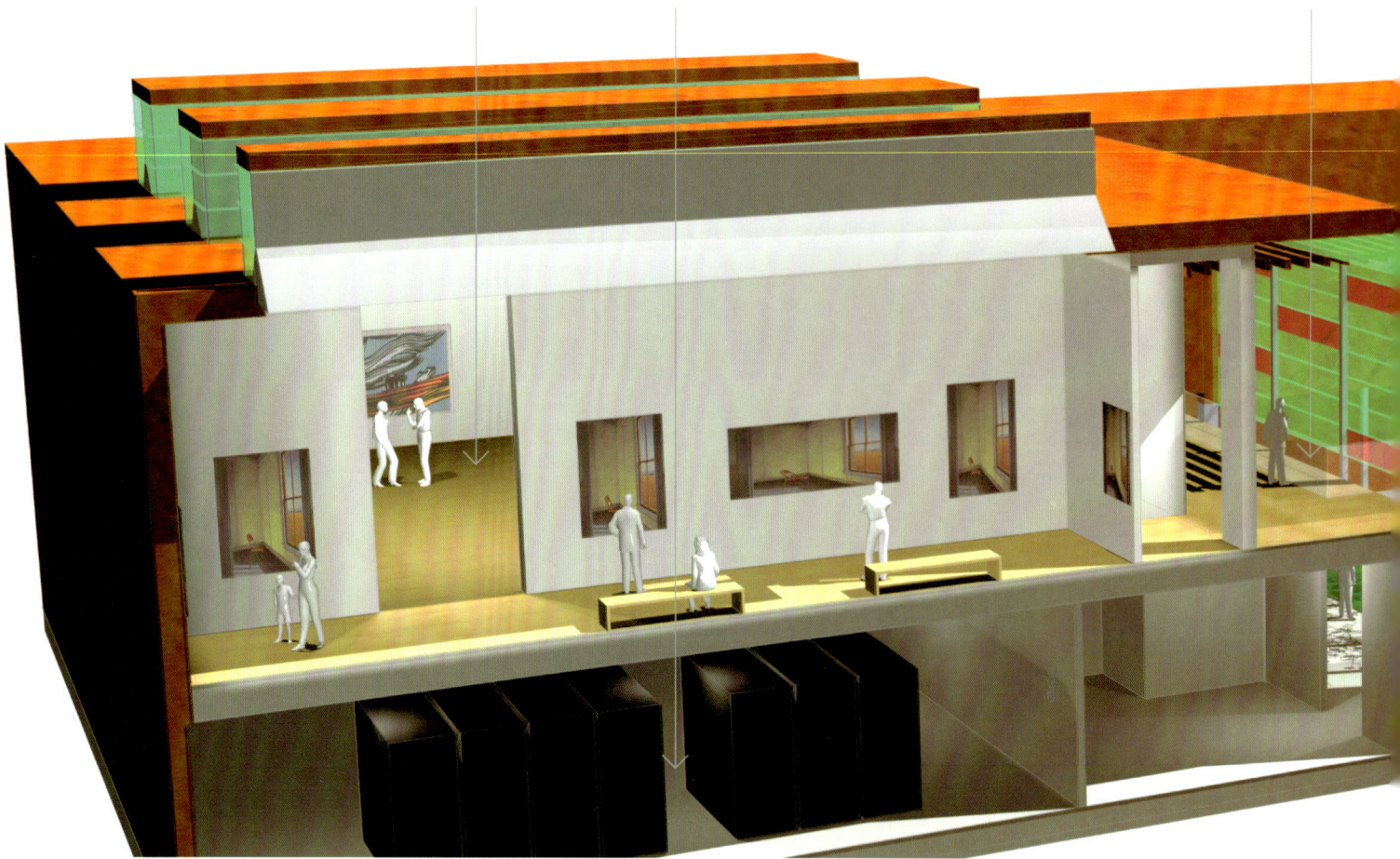

Section 剖面图

Children's play frame 儿童戏剧框架图

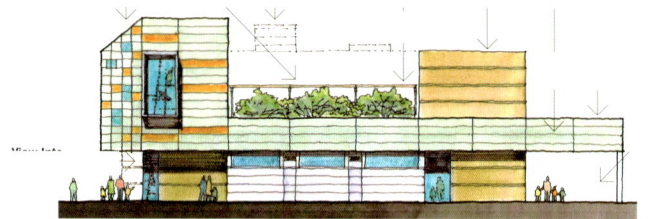

South elevation 南立面图

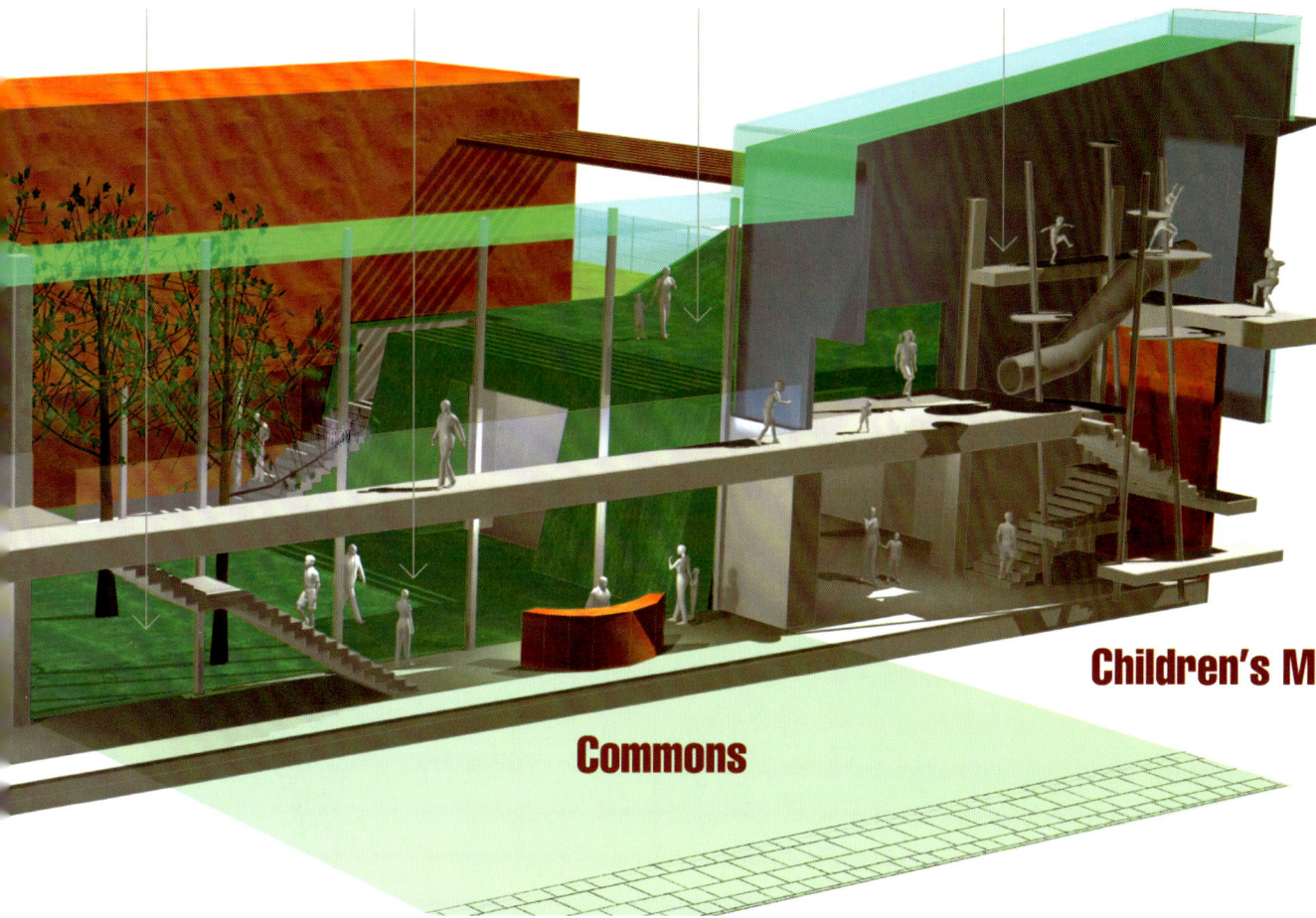

Commons

Children's Museum

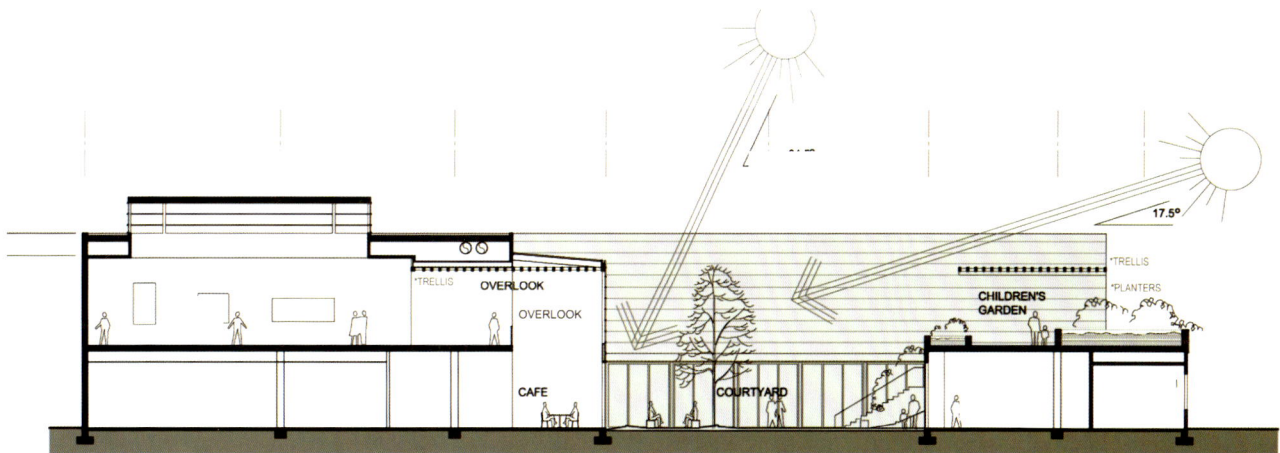

*TRELLIS   OVERLOOK

OVERLOOK

CAFE

COURTYARD

CHILDREN'S GARDEN

*TRELLIS

*PLANTERS

17.5°

Czech National Library Competition

捷克国家图书馆竞赛

竞赛背景 Competition Background

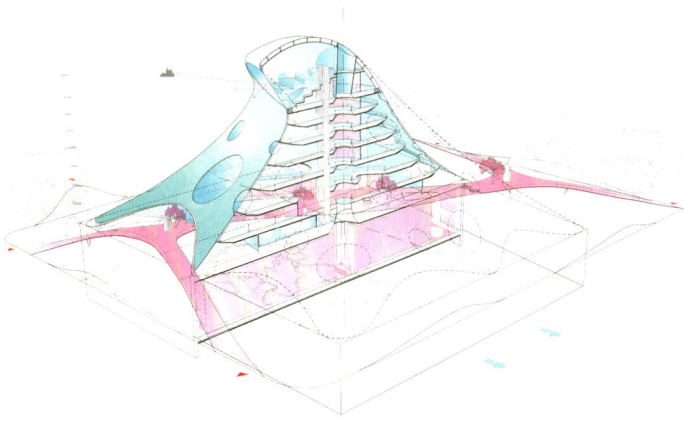

In the "Golden City", Prague, the Czech National Library is situated in the baroque complex of buildings known as the Clementinum. Technically at a mid-20th-century level, the library itself is hopelessly overcrowded and difficult to navigate. Readers report that it takes nearly three hours to obtain an ordered book. It currently houses 6,000,000 volumes with 80,000 volumes aded annually. Though it has stored its overflow in a space in the outlying district of Hostivar and digitized selected stock since 1997, the lack of space continues to be a problem.

After half a century of discussion, the Prague authorities sold a site to the state to build a new facility in the early 2006. The site is located on the Letensk sady (Letna plain) and overlooking the Moldau from the north. Now it is occupied by a small bus and tram station.

在"金色之城"布拉格，捷克国家图书馆坐落在一片名为"克莱门提纳姆"的巴洛克建筑群。在20世纪中叶，图书馆的藏书书满为患，问题很难解决。读者们抱怨要将近三个小时才能拿到预定的书。目前，图书馆藏书600万册，而且还在以每年8万册的速度增加。尽管它从1997年起，就将过剩的图书存储在偏远的豪斯提瓦区，并采取了数字选书系统，缺乏空间的问题一直没有得以解决。

在半个世纪的讨论后，2006年初，布拉格当局出售给国家一块土地新建新图书馆。场地坐落在莱特纳平原，从北侧俯视莫尔道河，场地上现有一座小型公交和有轨电车站。

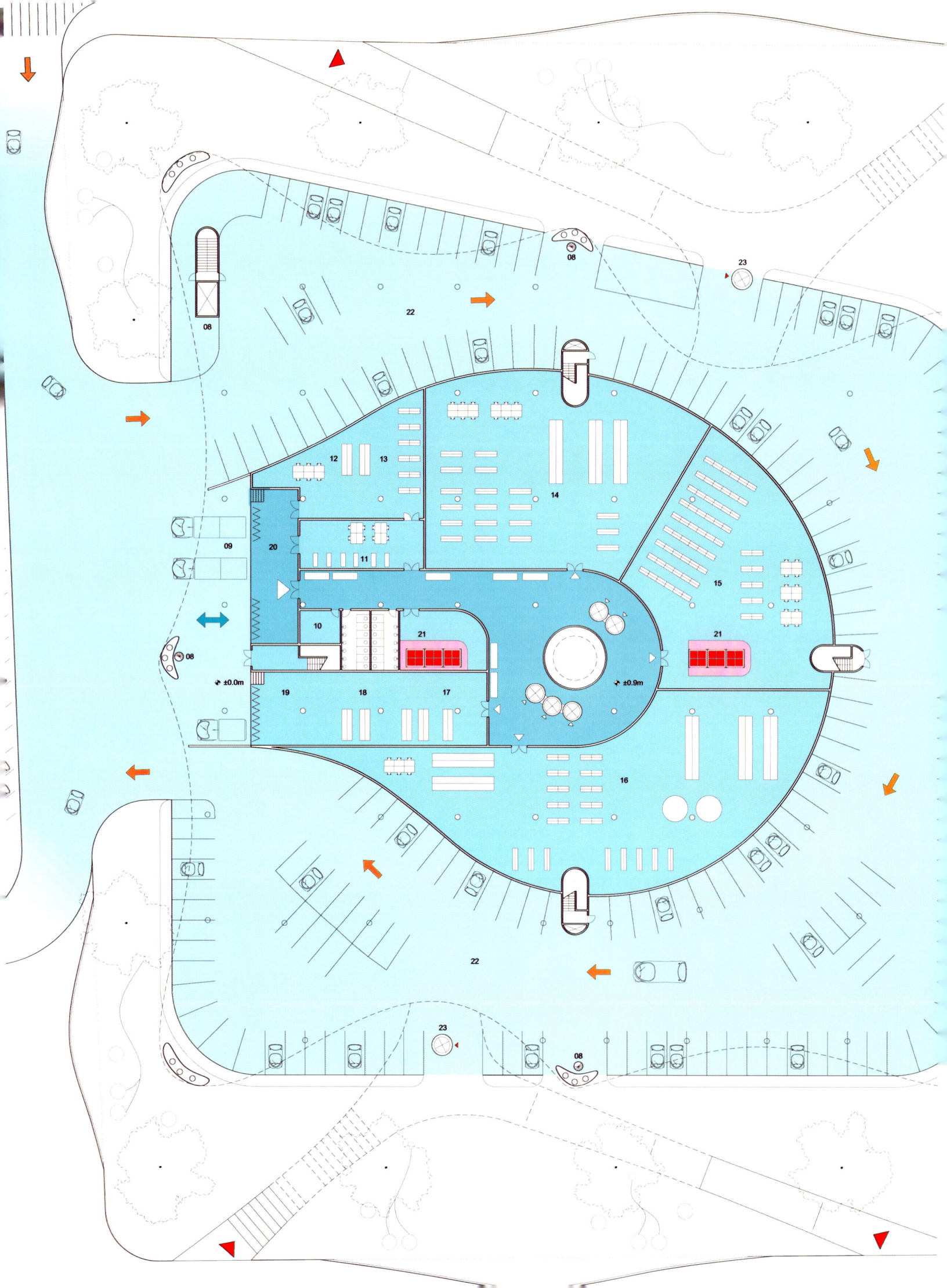

08

23

22

08

12 13

14

09 20

11

15

±0.0m 10

08 21 21

19 18 17 ±0.9m

16

22

23

08

专家点评   **Remarks**

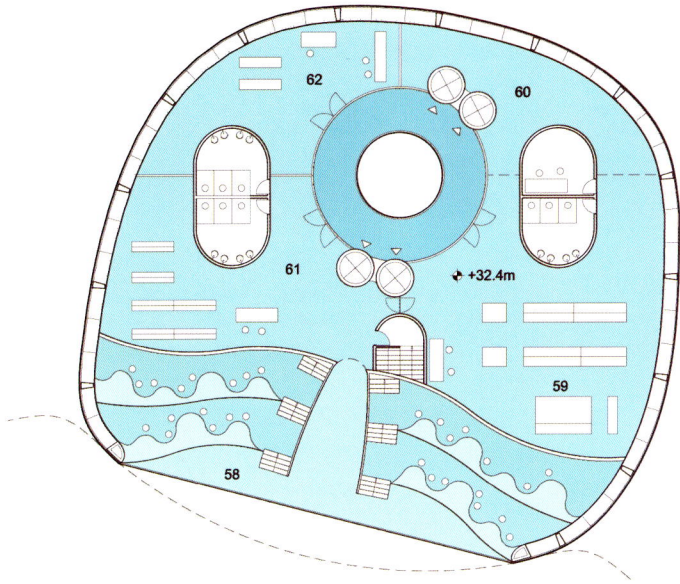

As to the winning entry's form, the jury was more than enthusiastic: "The mass growing smoothly from Letna plain does not represent a disturbing but rather a natural element, which by its shape evokes the impression of a natural component of a landscape (a hill)."

评审委员会对获奖设计的外观赞不绝口："主体结构从莱特纳平原缓缓升起，对环境完全没有破坏，反而更像一个自然的元素。它的形状让人想起了山脉的景观。"

未来系统公司（简称FS）
参赛项目1（一等奖）

Future Systems (FS) Project 1 the first prize

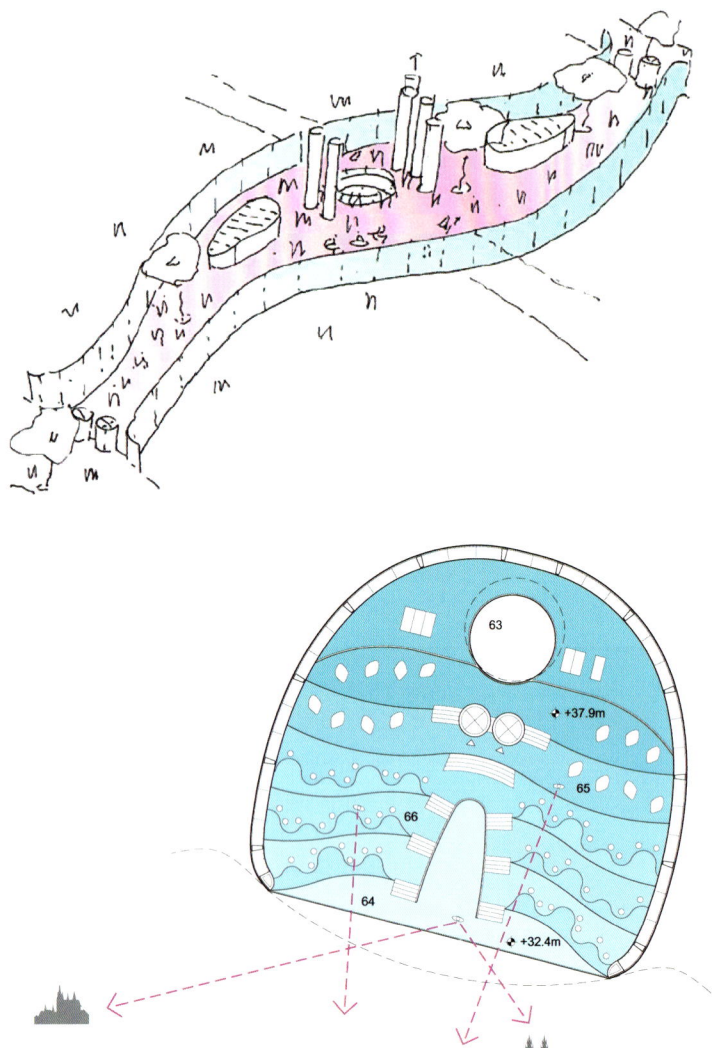

Although the design by Future Systems (FS) did not meet all requirements — they include only 40,000 of the 50,000 square meters specified — its unusual configuration carried the day in the ajudication process. Books and bookshelves were the only right angles in the building. The design shows an amorphous structure devoid of a single straight line on the exterior. FS developed the form and curvature in reference to baroque buildings in Prague, creating a champagne-colored anodized-aluminum tile skin fades from dark to light from bottom to top. According to FS, the building will be placed on a white unpolished marble platform, with mirror-finished stainless rings around the perimeter to reflect the 66-meter-high building from different angles.

Below the street level, the stacks stand 18 meters high and occupy the entire 11,000 square meters. The distribution of the estimated 10 million books stored there will be facilitated by an Automated Storage and Retrieval System and reach the reader in less than five minutes. Entrances for readers and ramps for cars and delivery vans surround the street level. Ramps and stairs from the outside lead to an interior walk, lined with trees and benches, which weaves through the nine-story building.

Brightly colored reading rooms are interspersed through these floors. Circular areas of glazing distributed over the exterior provide generous levels of natural light in all public spaces. A coffee house occupies the top level where ceiling are 37.9 meters high as opposed to the 4.5 meters on the levels below. The stepped-up floor alows café patrons to sit in undulating rows overlooking Old Town and the Castle through a large picture window to the south/southwest.

尽管未来系统公司（简称FS）的设计并未完全符合要求——竞赛要求建筑面积为5万平方米，而这个项目只达到了4万平方米，它不寻常的造型使它获得了评审环节的胜利。书和书架是建筑里唯一的直角。设计采取无固定形态结构，外部没有一条直线。FS设计的外观和曲线借鉴了布拉格的巴洛克建筑，打造了一个香槟色阳极化铝外层物料，颜色由下至上，逐渐变浅。FS称，整个建筑将放置在一个白色未抛光的大理石平台上，周边是镜面的不锈钢圆环，可以映出这幢66高的建筑的各个角度。

地下的多层书架高18米，所占面积为1.1万平方米。储藏在那里的1000万本图书将由自动配对仓储和检索系统调配，读者将在5分钟之内拿到自己想要的书。街面四周是读者入口和车辆运输坡道。坡道和台阶通往一条室内人行道，人行道两侧是树木和长椅，蜿蜒的穿过九层的图书馆。

色彩明亮的阅览室点缀着各个楼层。建筑外墙上的圆形嵌装玻璃为公共空间提供了大量的自然采光。咖啡厅设在顶楼。和其他高为4.5米的楼层不同，顶楼高37.9米。阶梯状的地板让喝咖啡的读者可以在起伏的座椅上从面向南方的窗子俯瞰老城和城堡的景色。

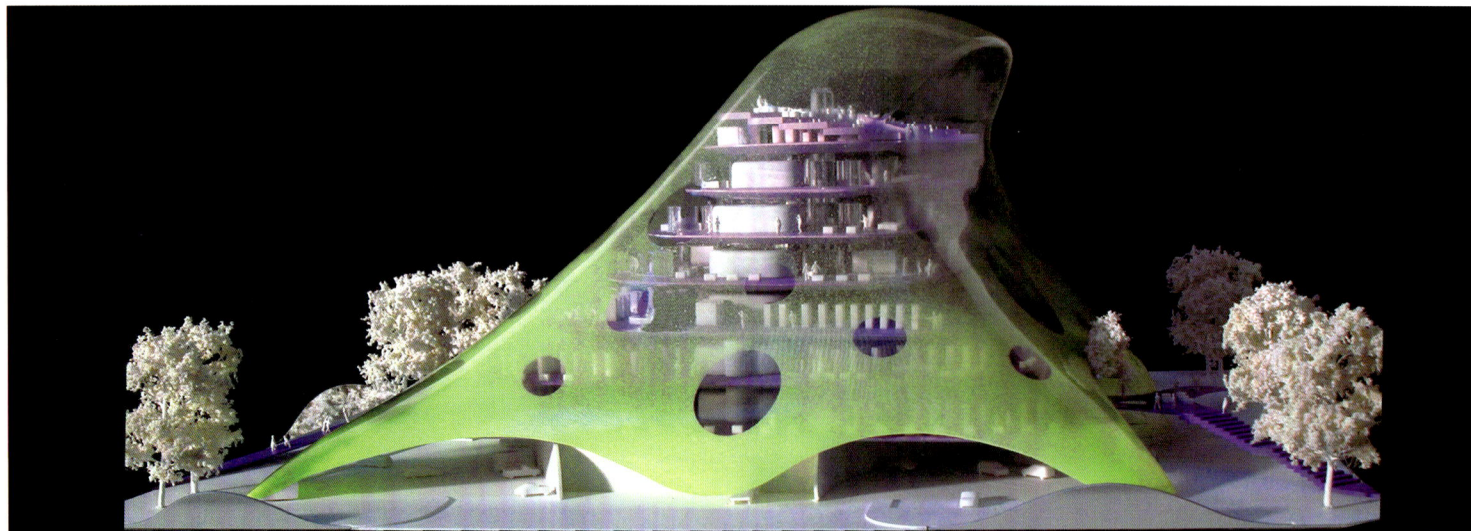

Model 模型

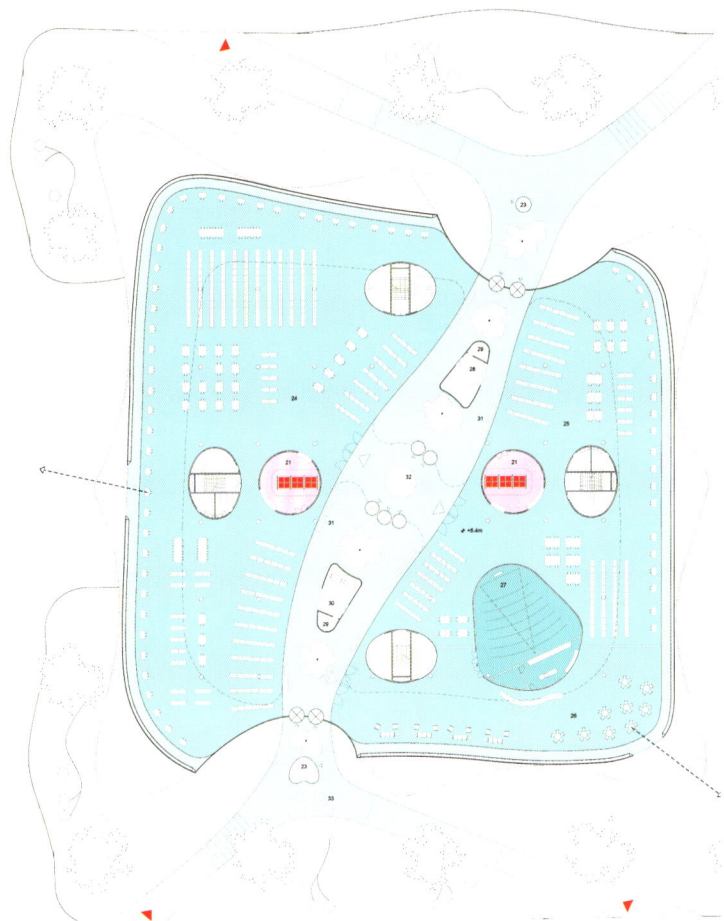

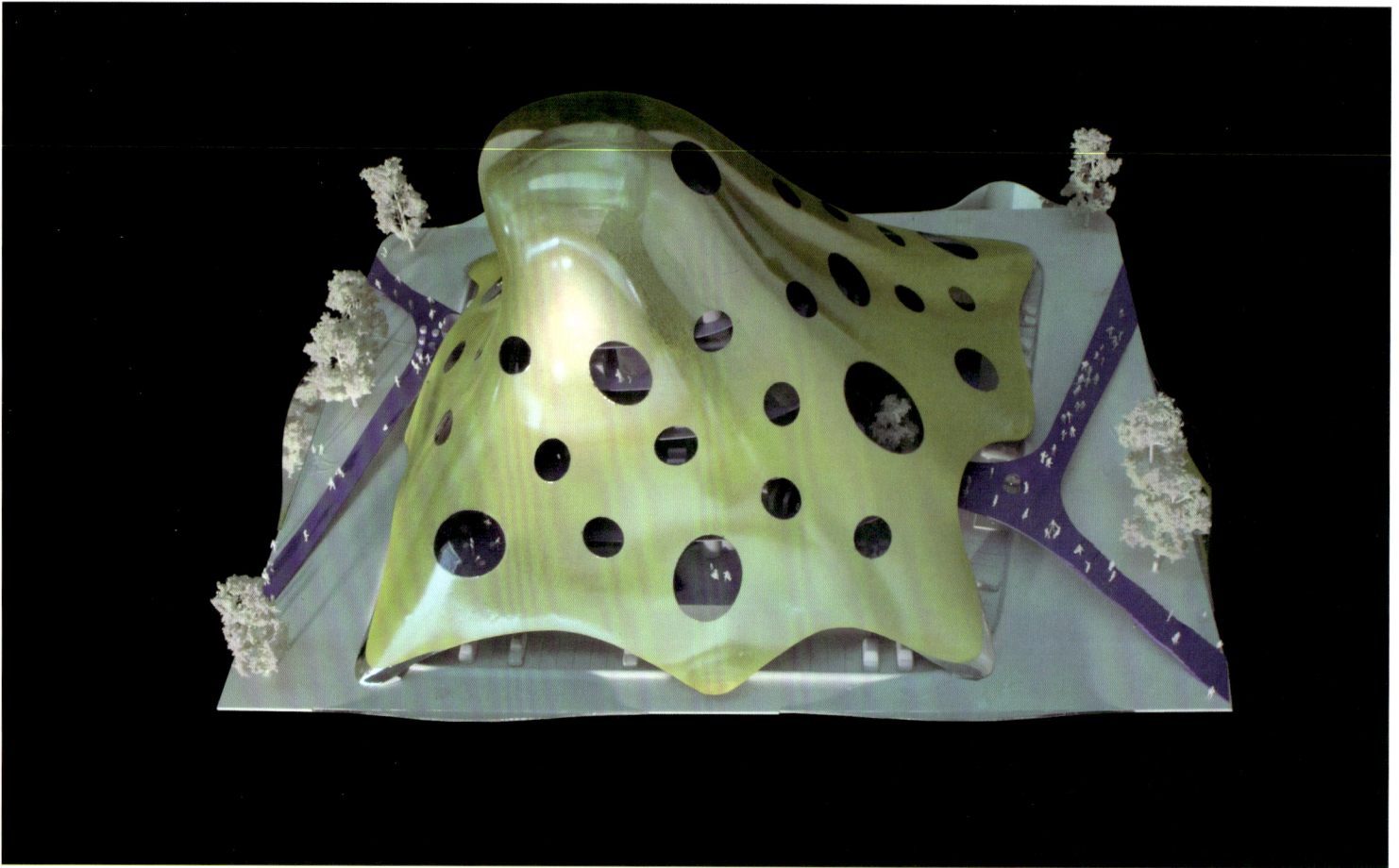

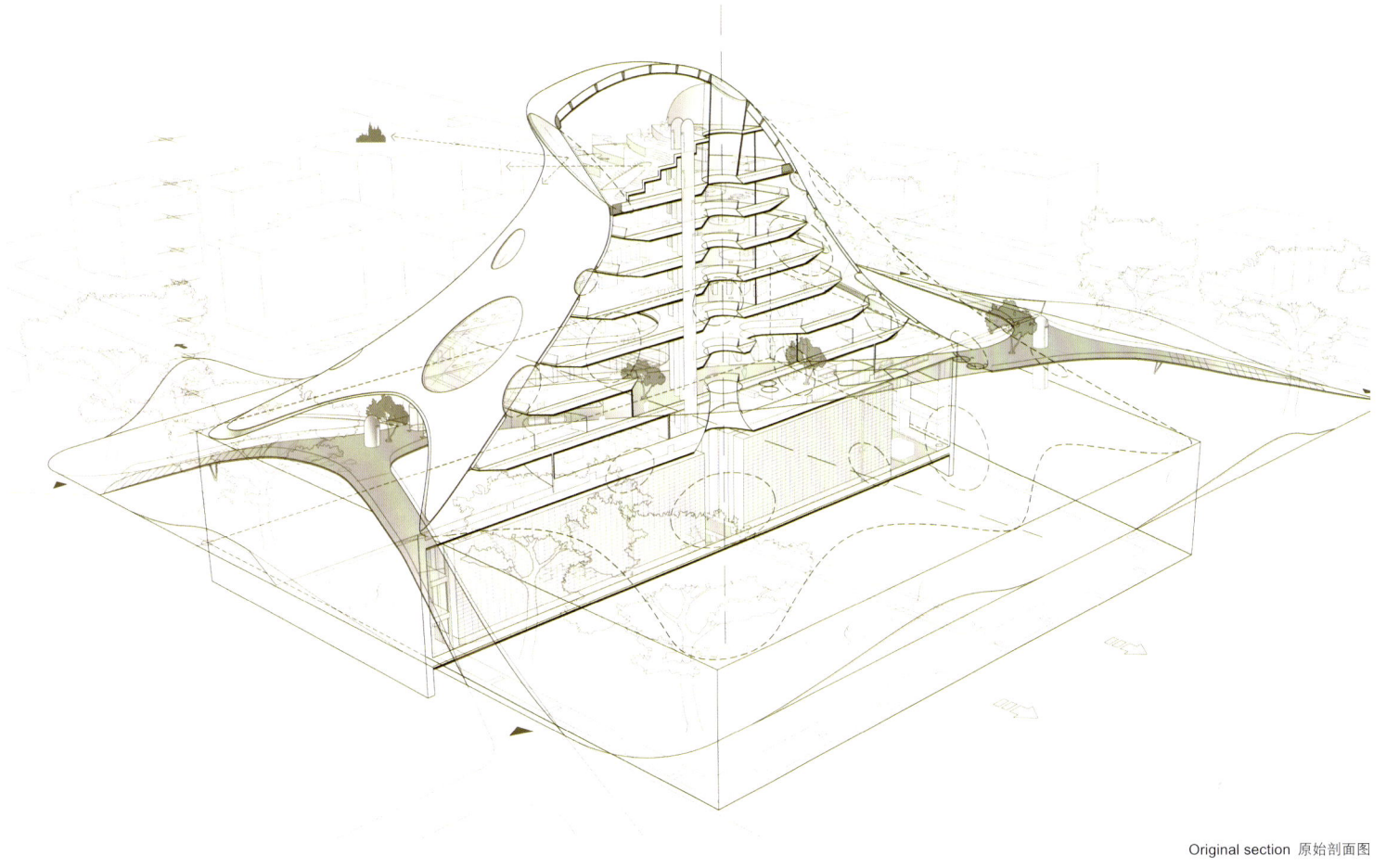

Original section 原始剖面图

Interior street view 室内街道

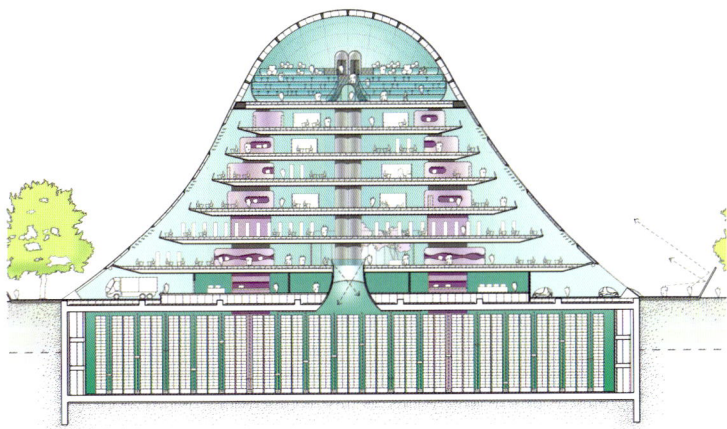

Section 剖面图

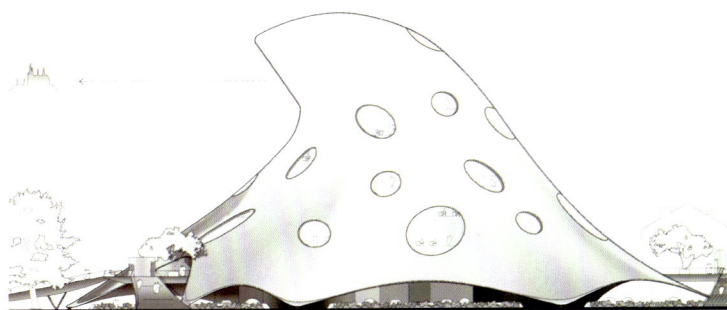

卡莫迪+格洛尔可 参赛项目2（二等奖） **Carmody Groarke Project 2 the second prize**

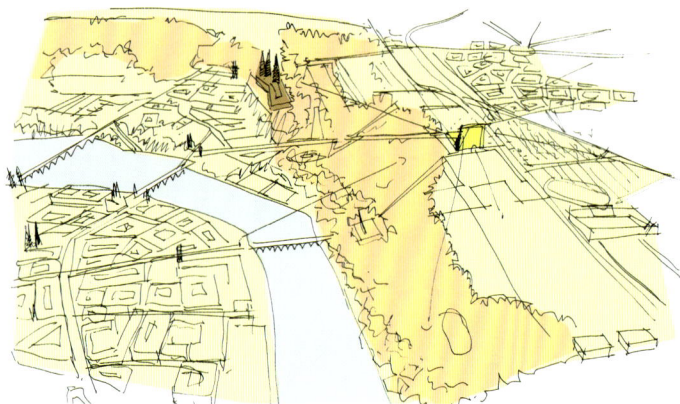

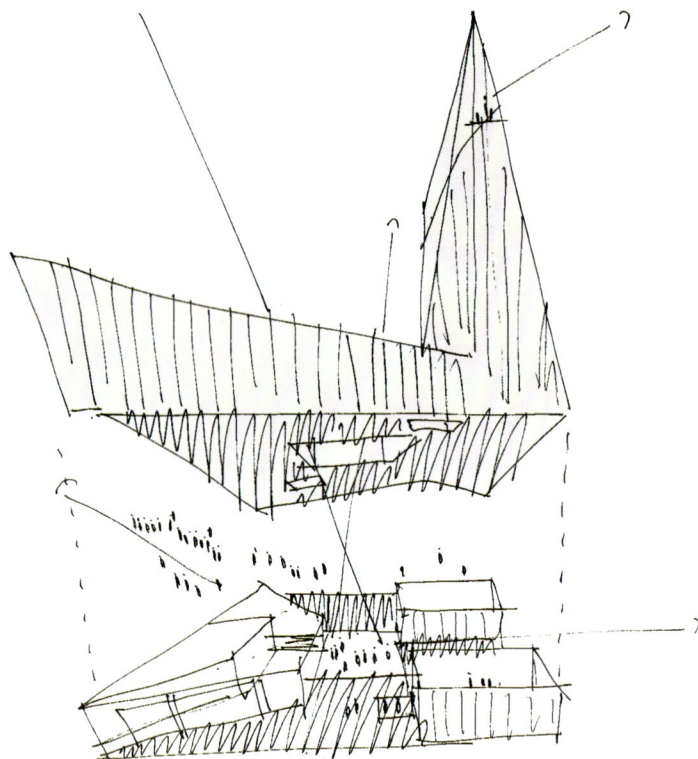

Book hall 藏书大厅

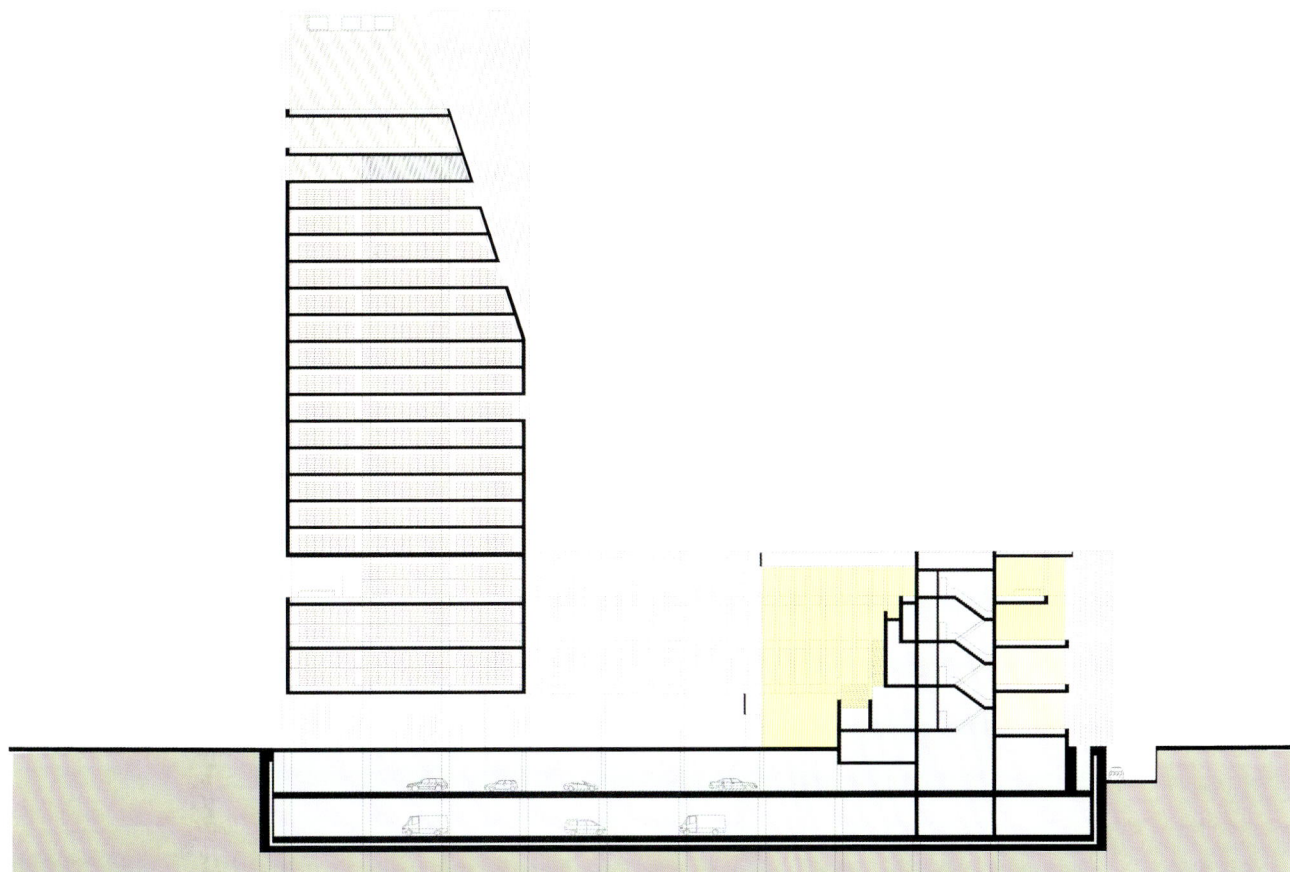

Library foyer 门厅

View to information desk 信息查询处

HSH建筑事务所 参赛项目3（三等奖） **HSH architekti Project 3
the third prize**

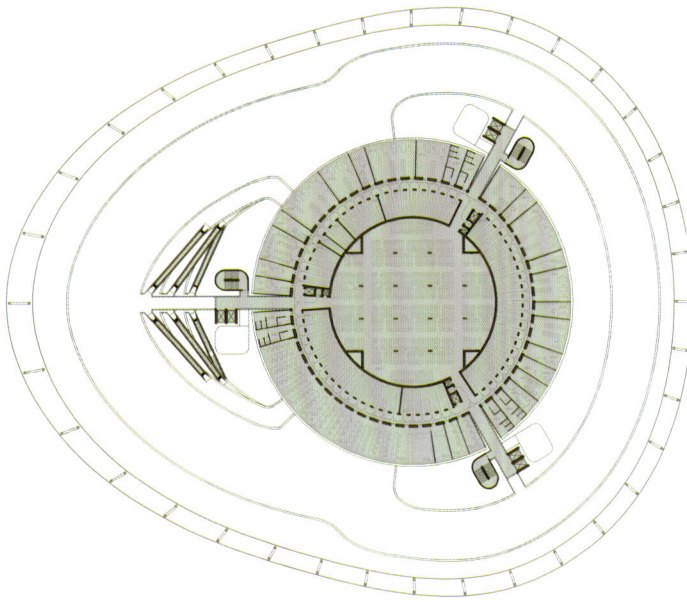 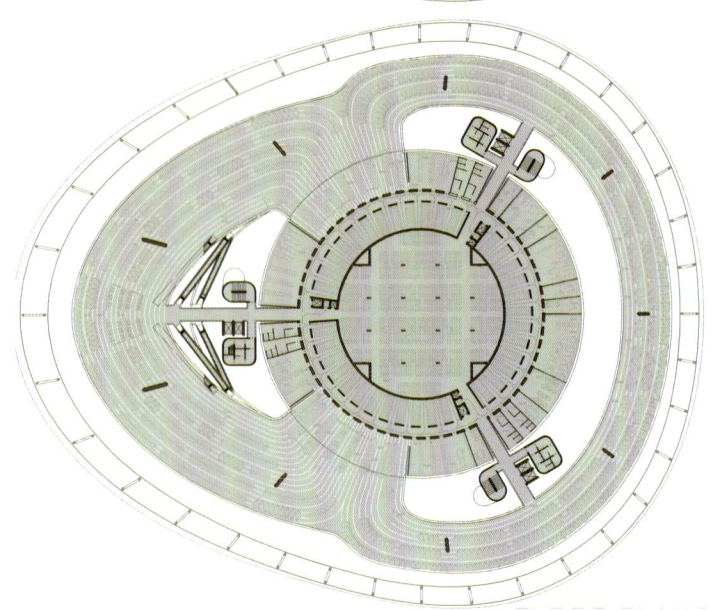

Aerial view of site 鸟瞰图

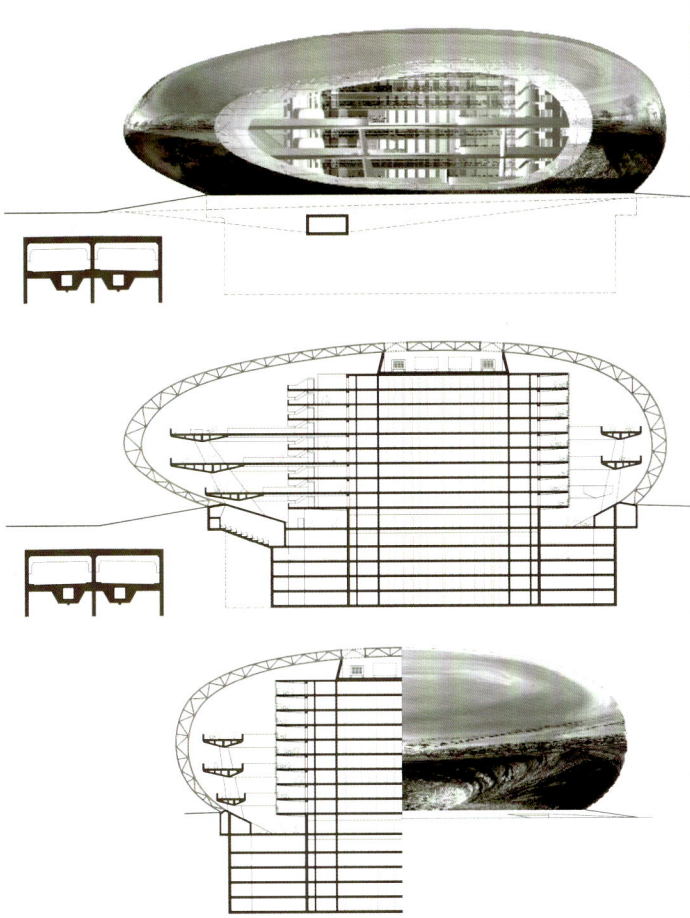

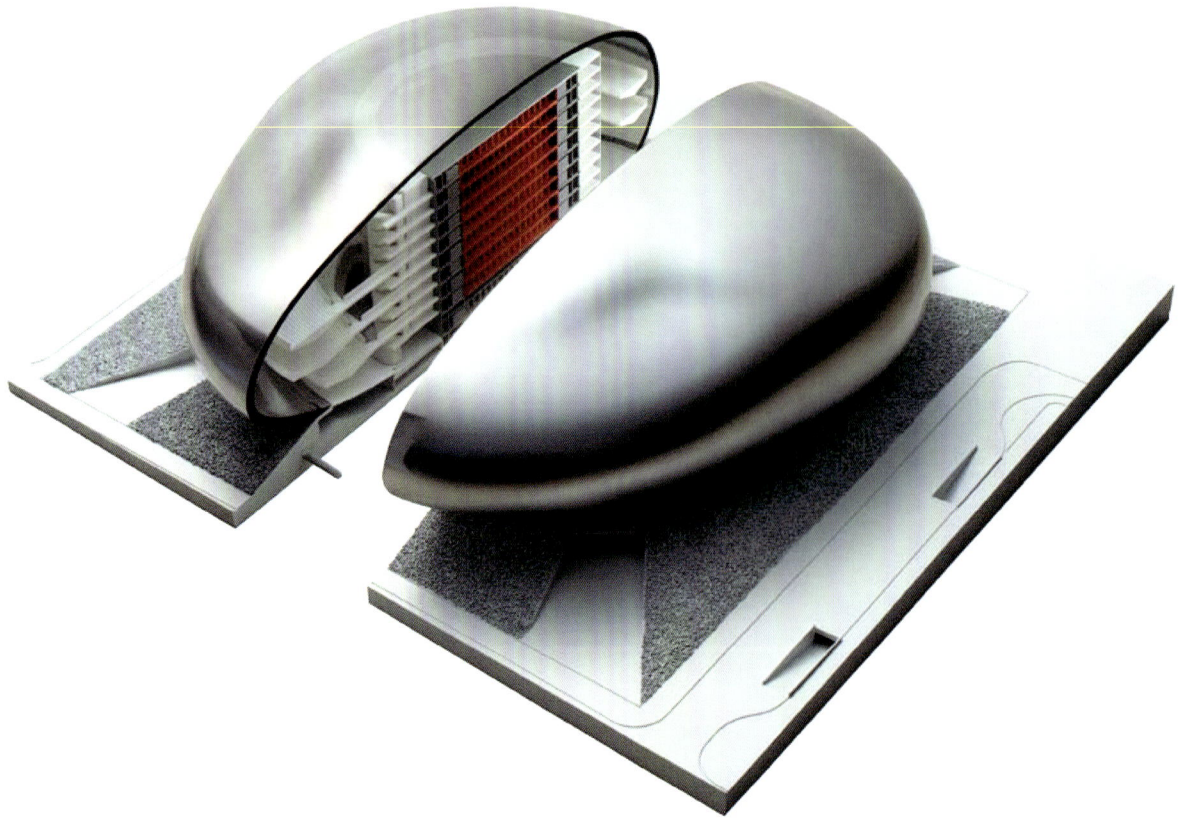

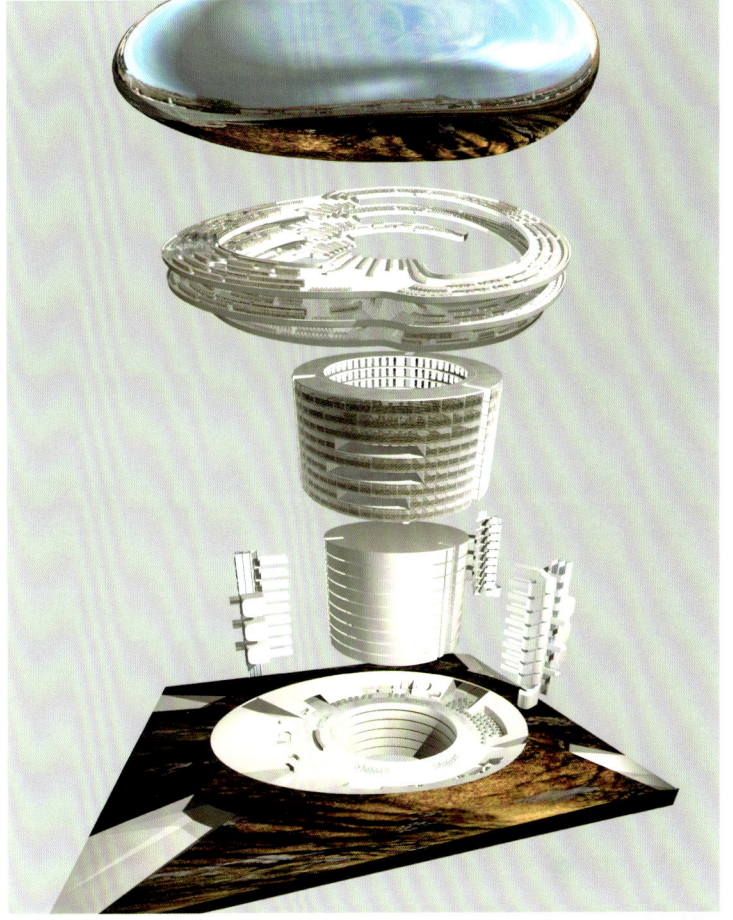

Exploded functional diagram 分解功能图

Interior perspective 内部景观

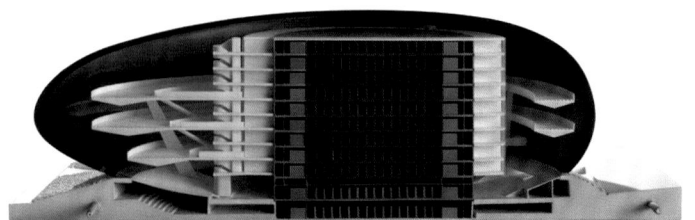

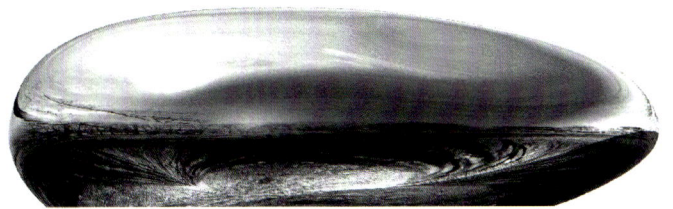

EMERGENT 参赛项目4（第四名）

EMERGENT Project 4
the fourth place

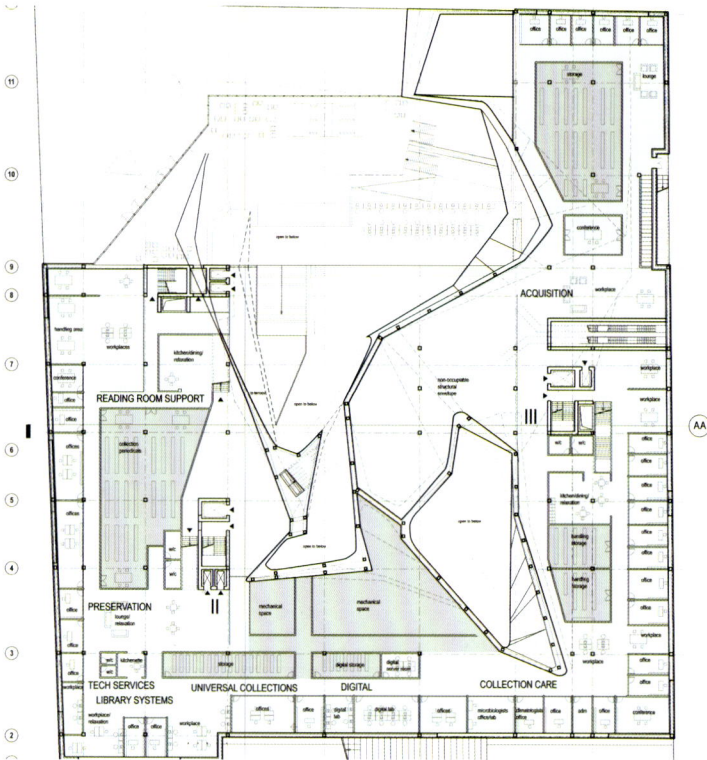

Night view 夜景

Interior perspective 内部景观

Interior 室内

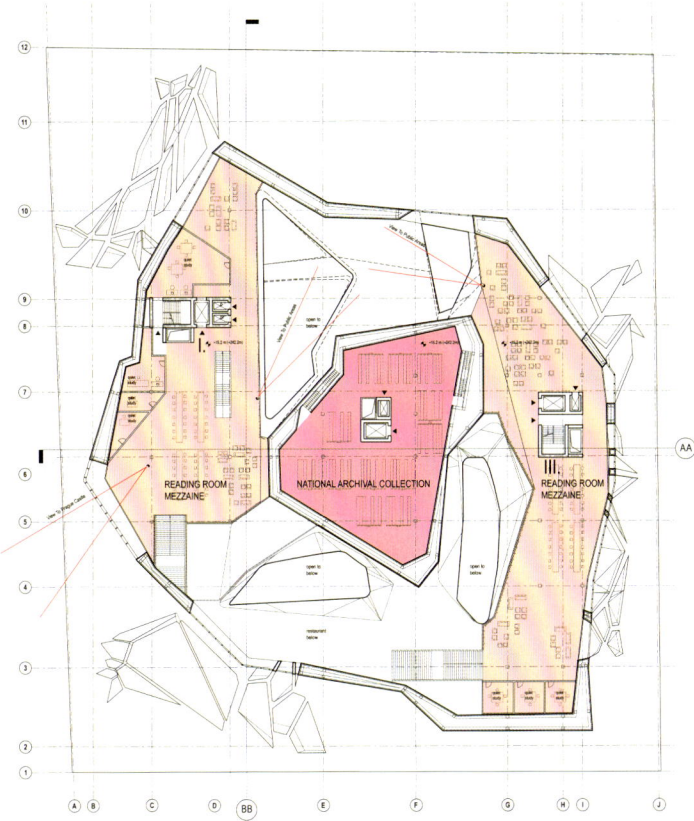

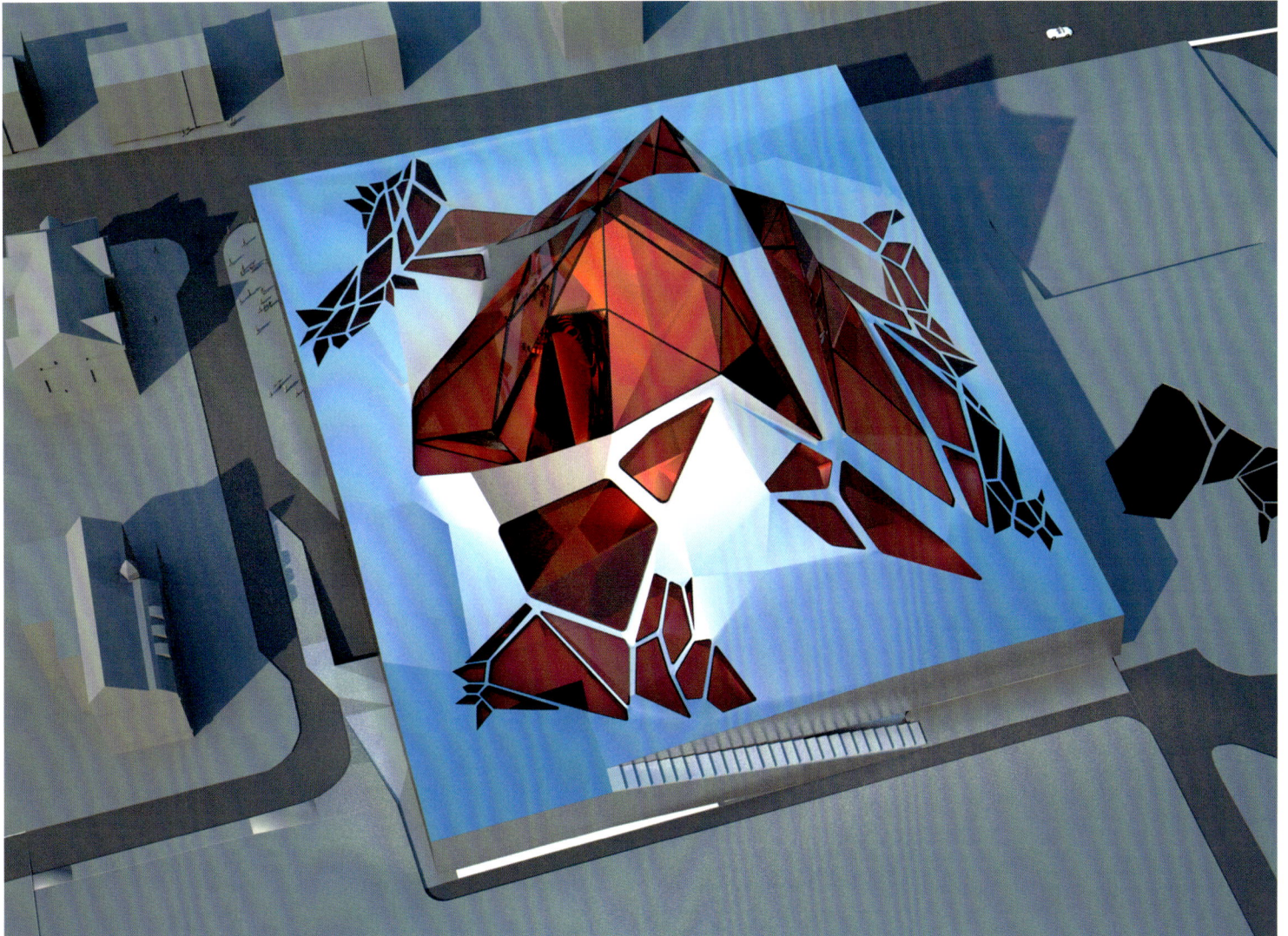

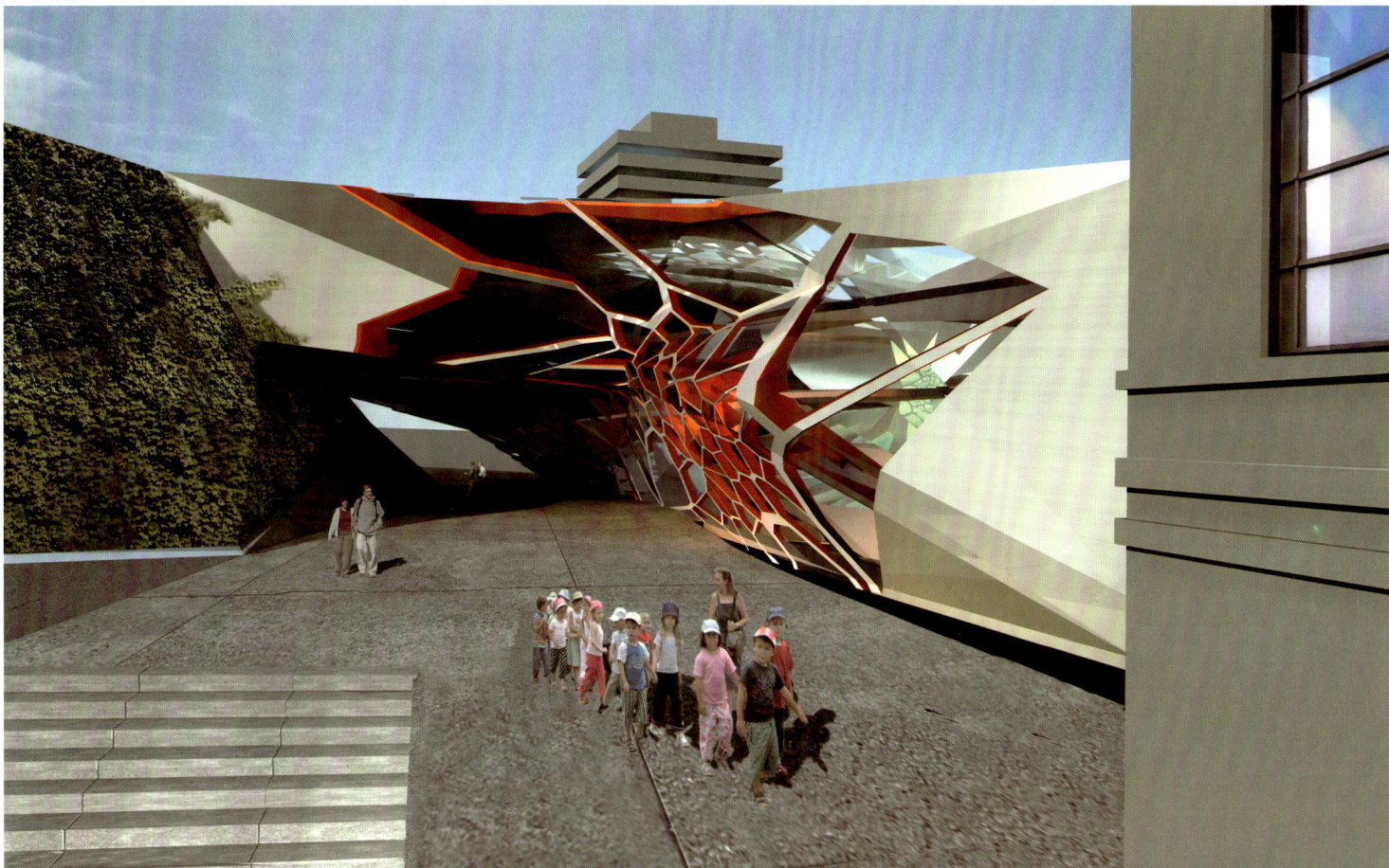

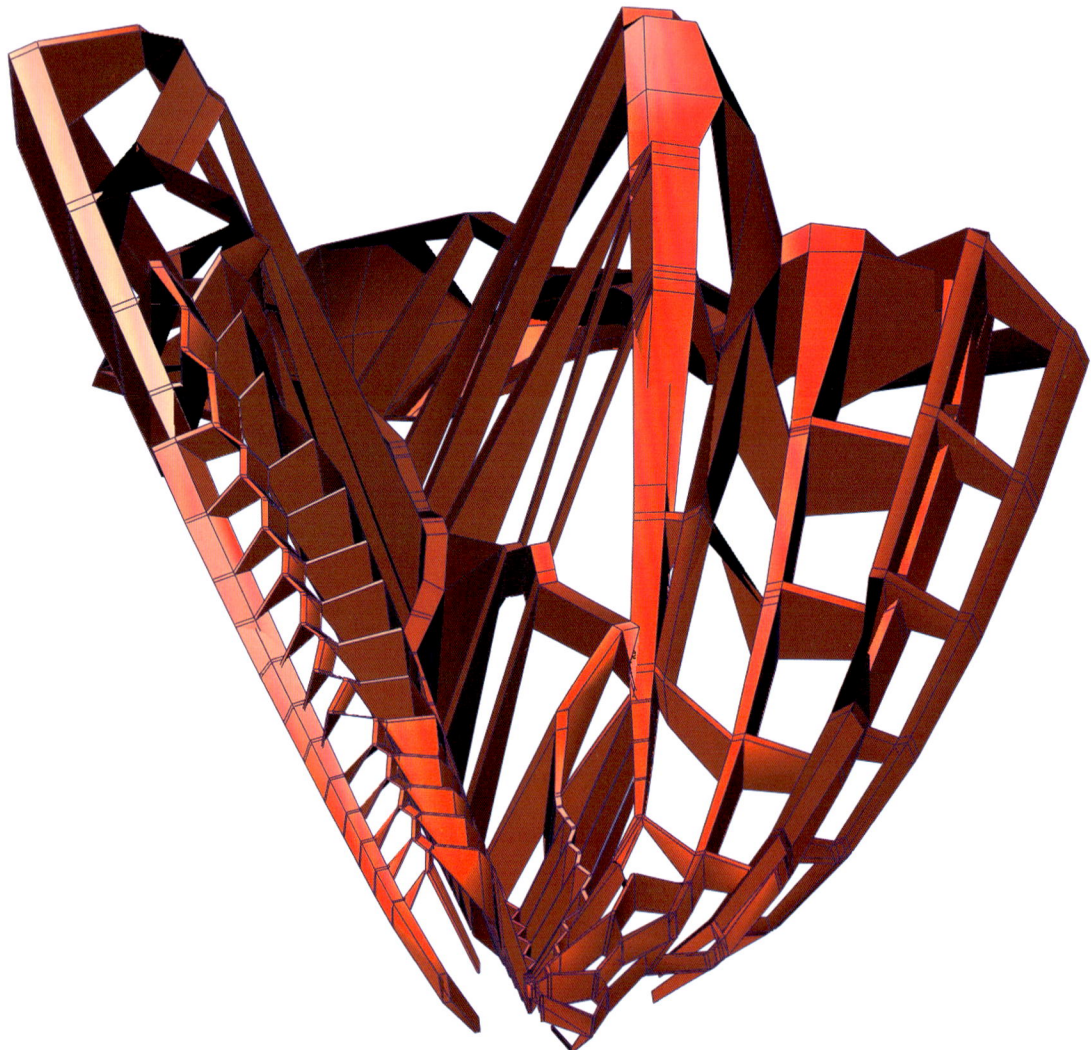

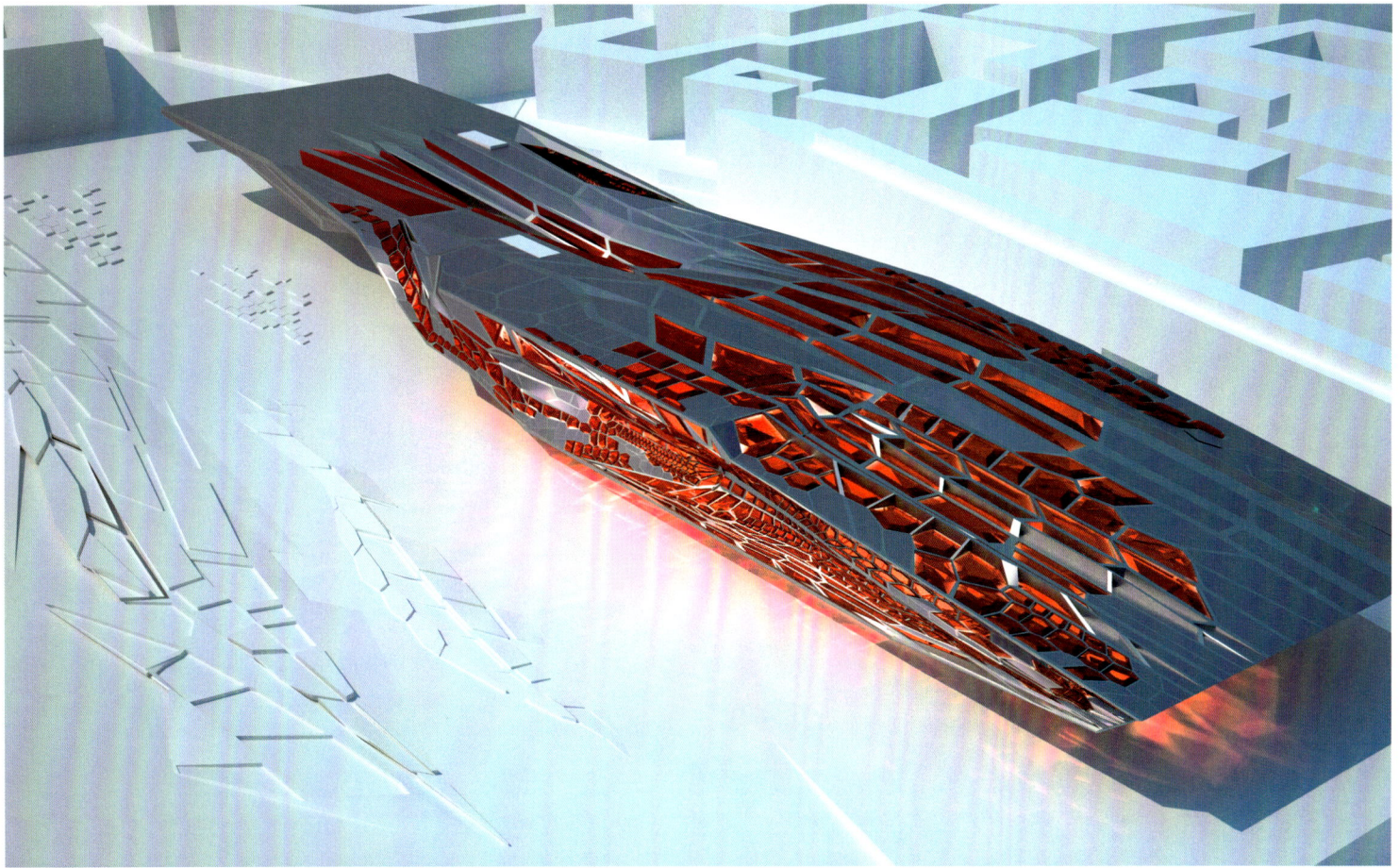

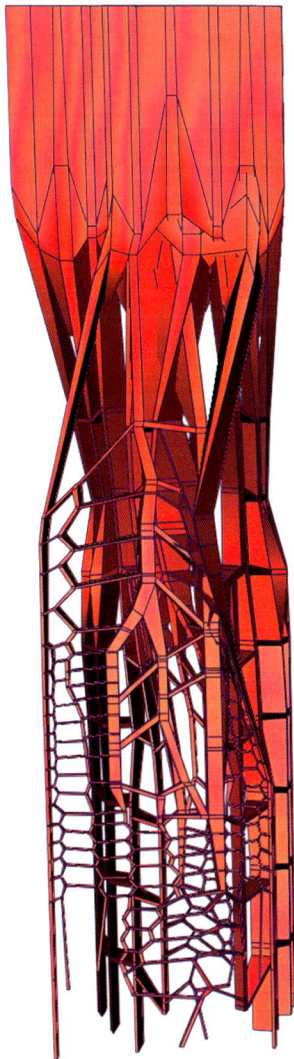

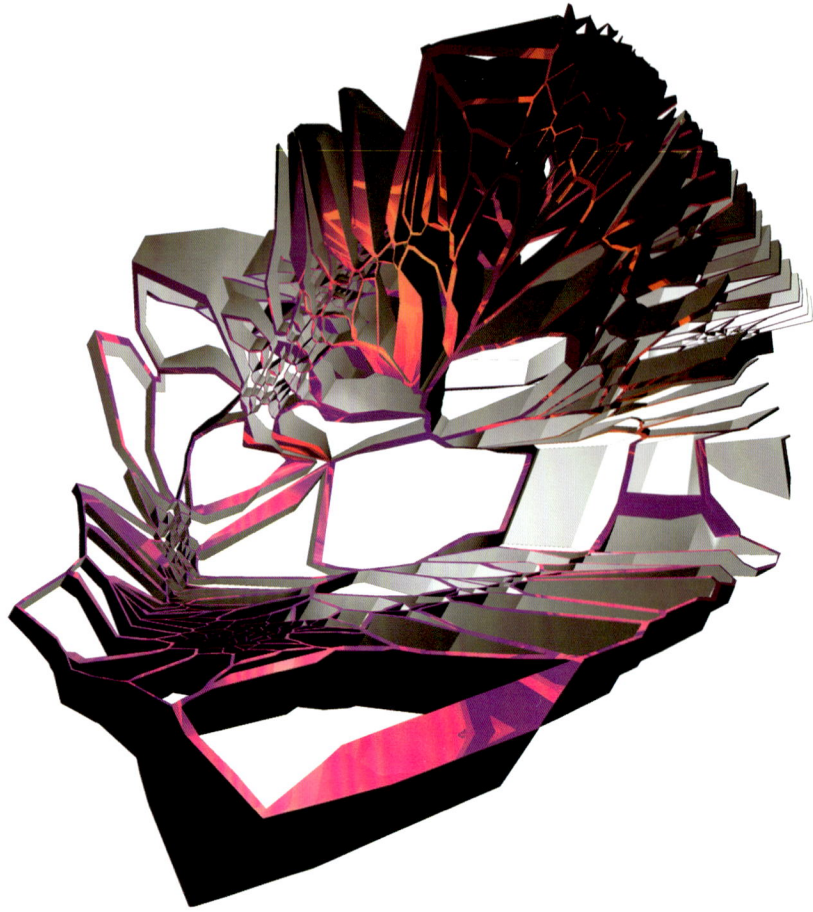

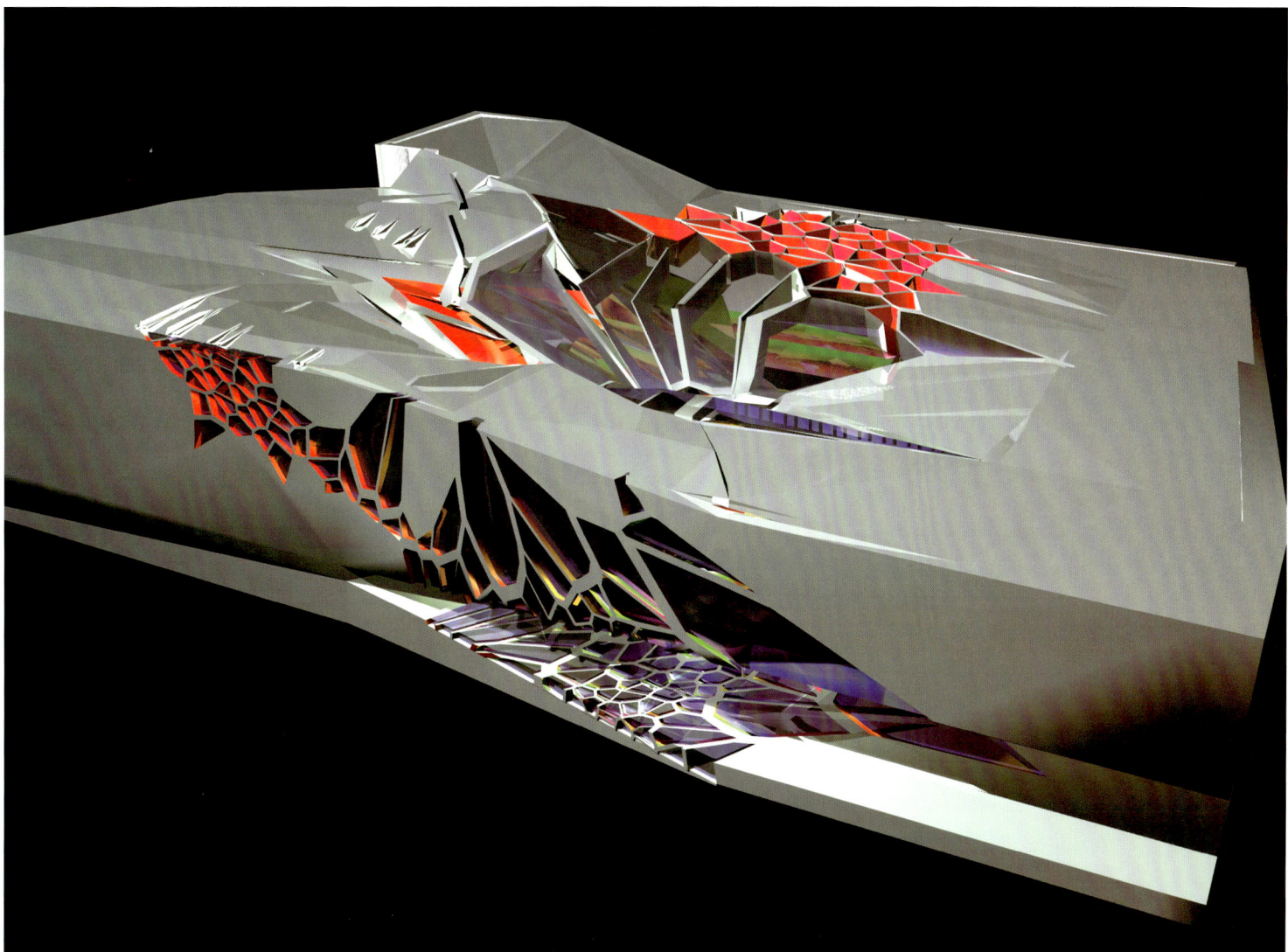

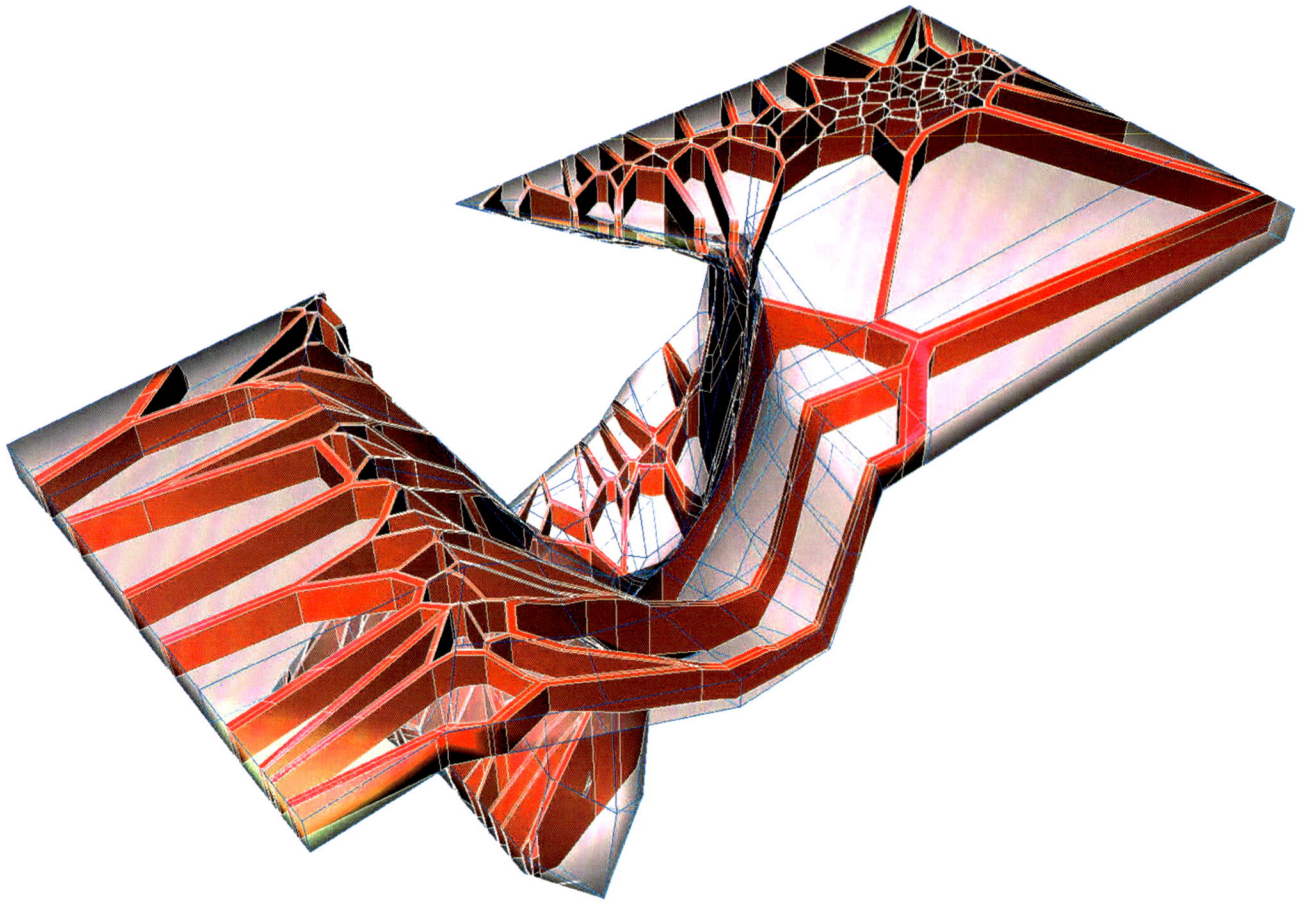

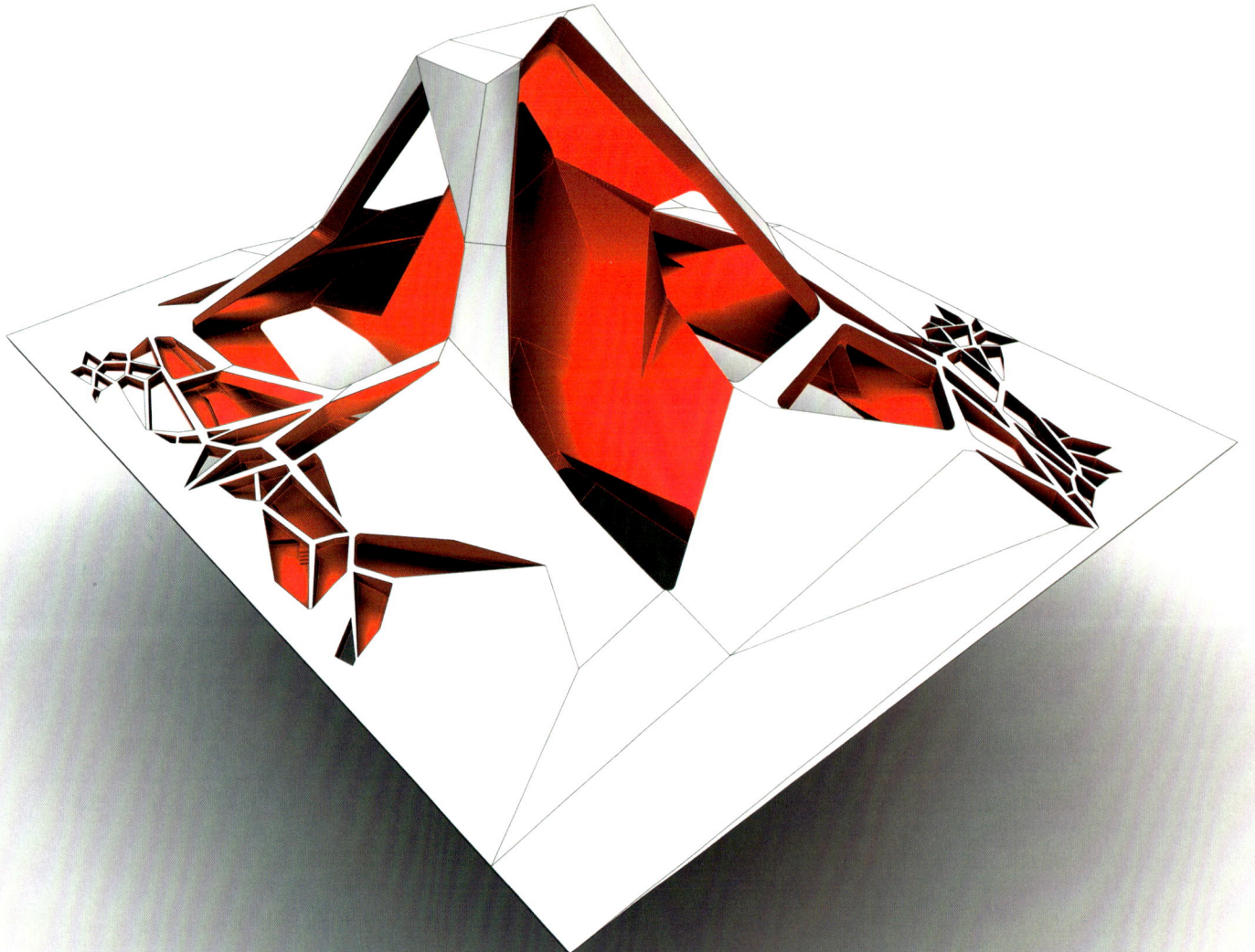

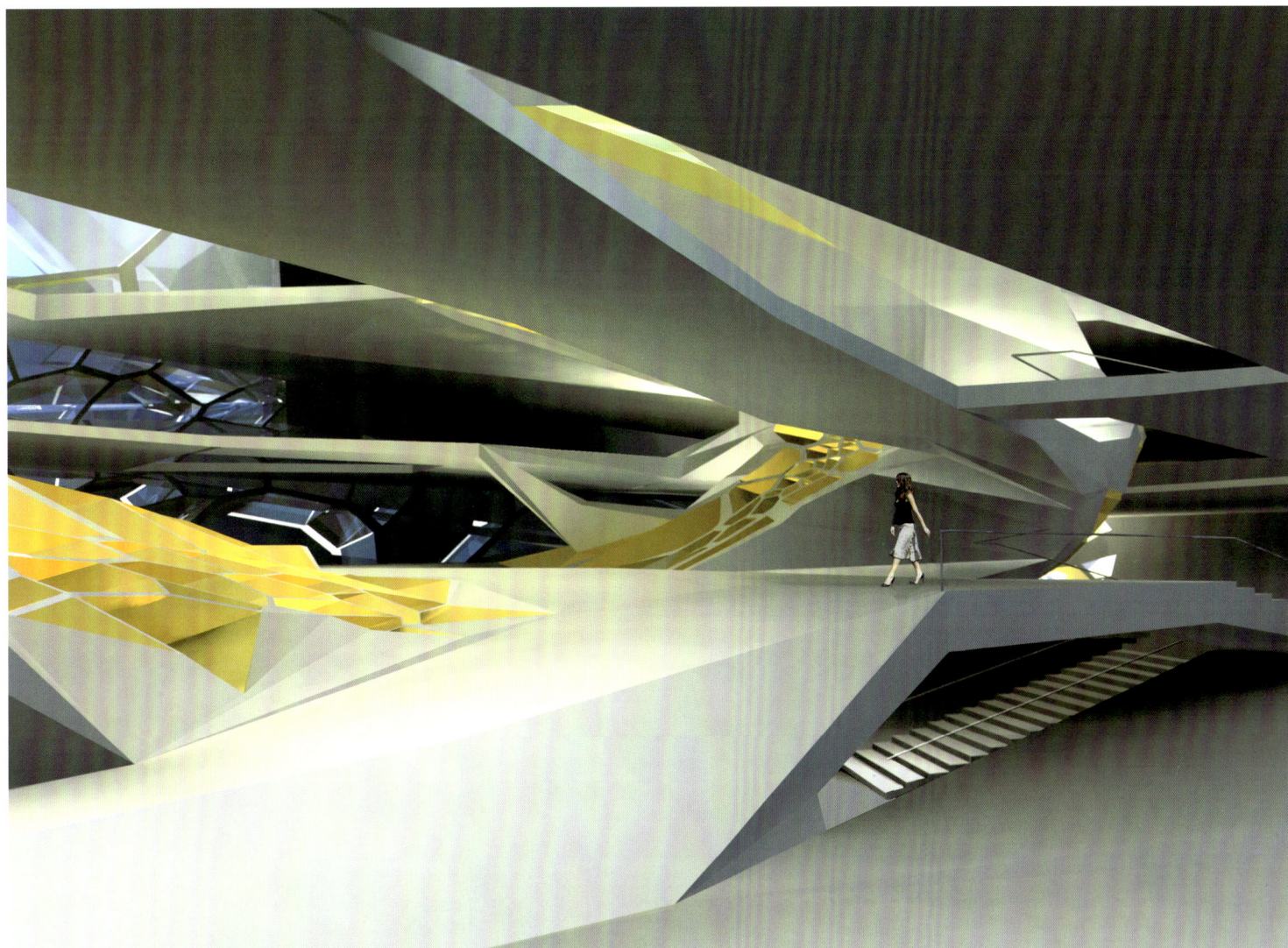

约翰·瑞德 参赛项目5（第五名）

**John Reed Project 5
the fifth place**

Night view 夜景

Night view to entrance 入口夜景

Interior view of foyer 大厅

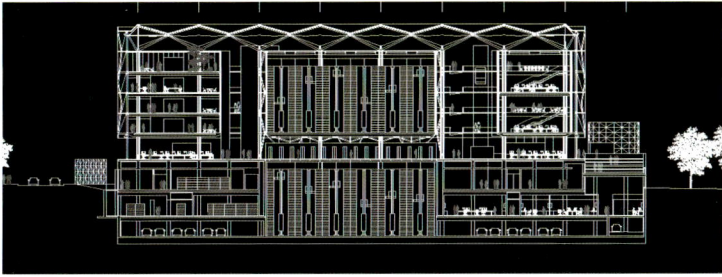

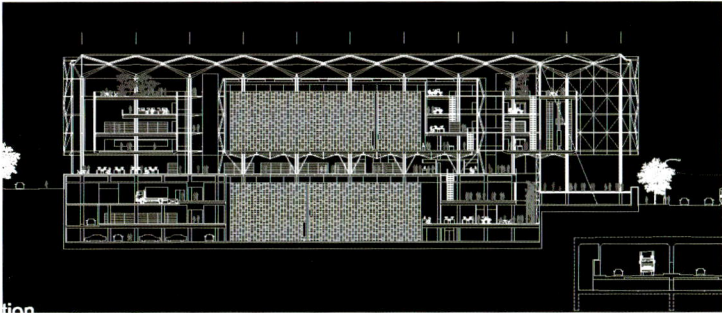

Sections 剖面图

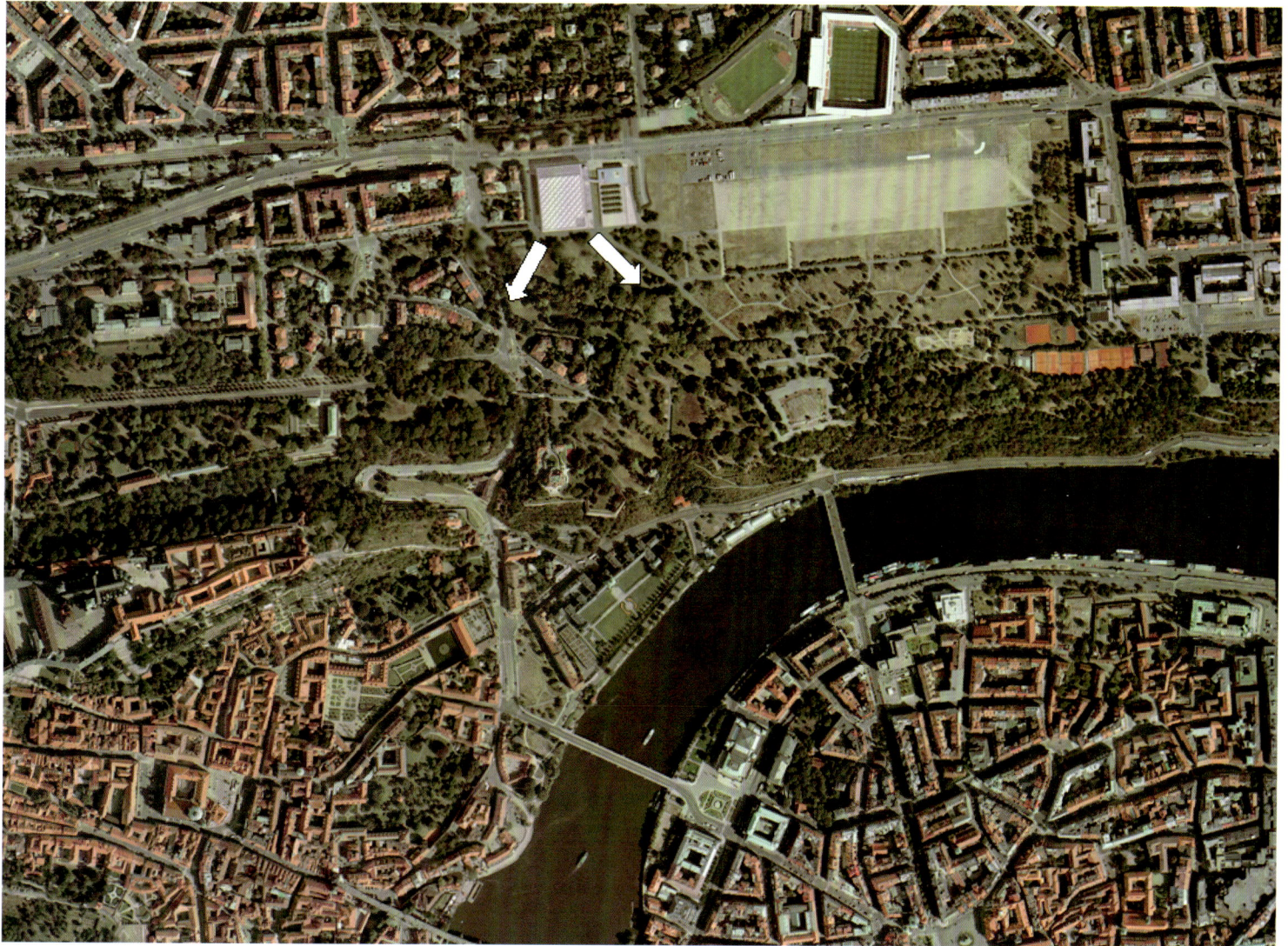

MVM建筑事务所+斯达克建筑事务所
参赛项目6（第六名）

MVMarchitekt + Starkearchitektur
Project 6 the sixth place

Stacks and reading area 藏书和阅览室

Stacks and reading area 藏书和阅览室

Night view 夜景

戴戈玛·理查特 参赛项目7（第七名）

Dagmar Richter Project 7
the seventh place

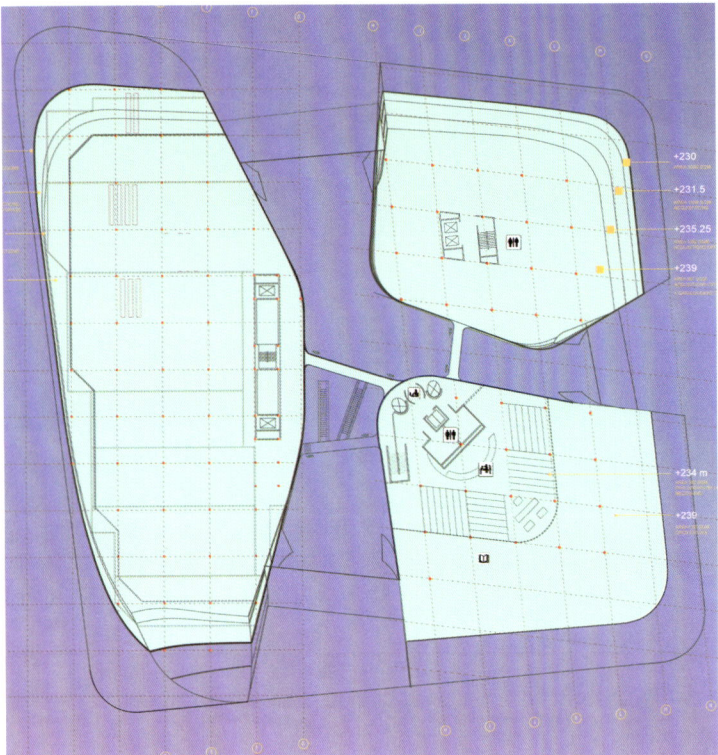

INTERIOR LATERAL BRACING

Model 模型

MODEL IMAGE

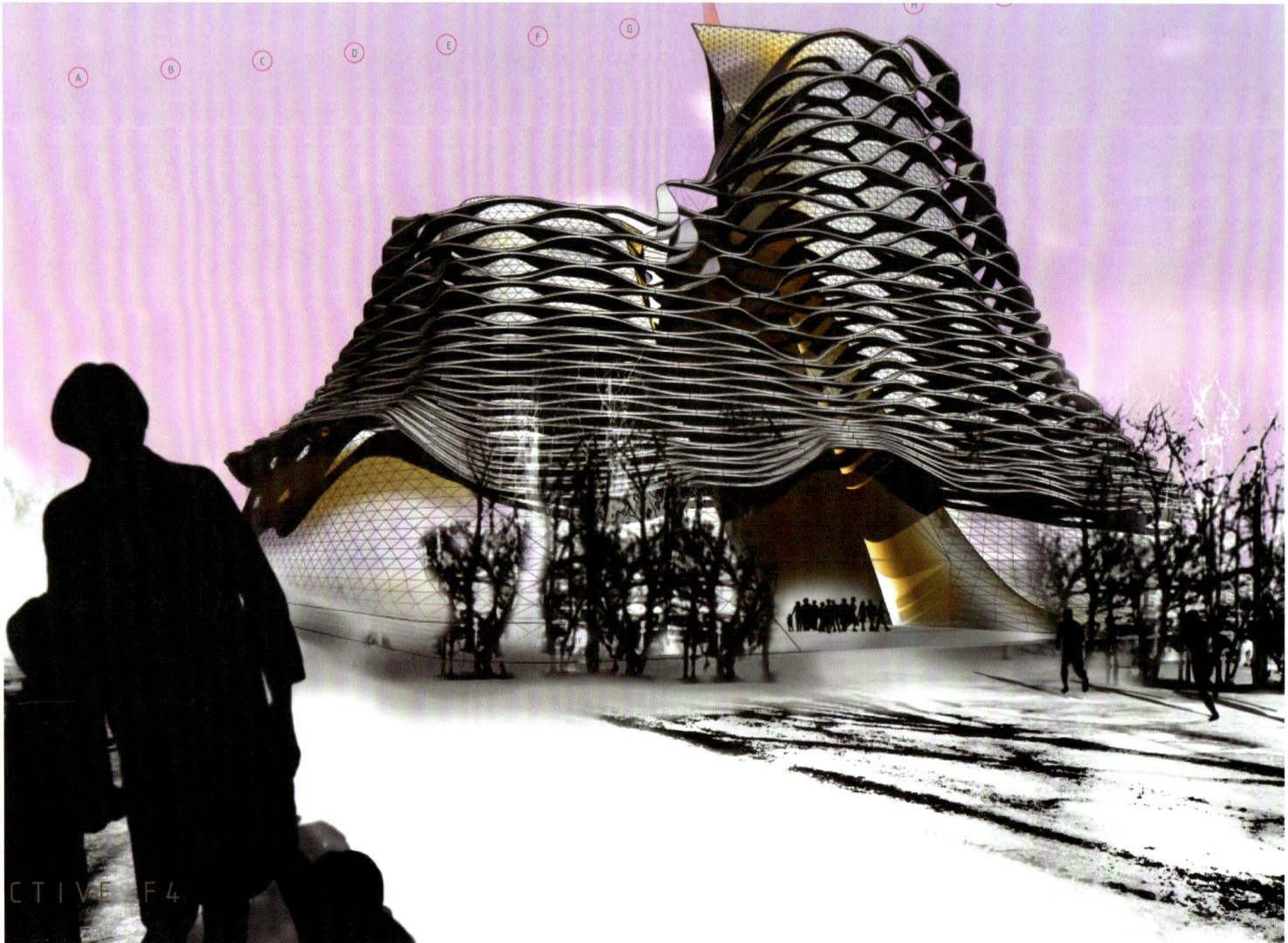

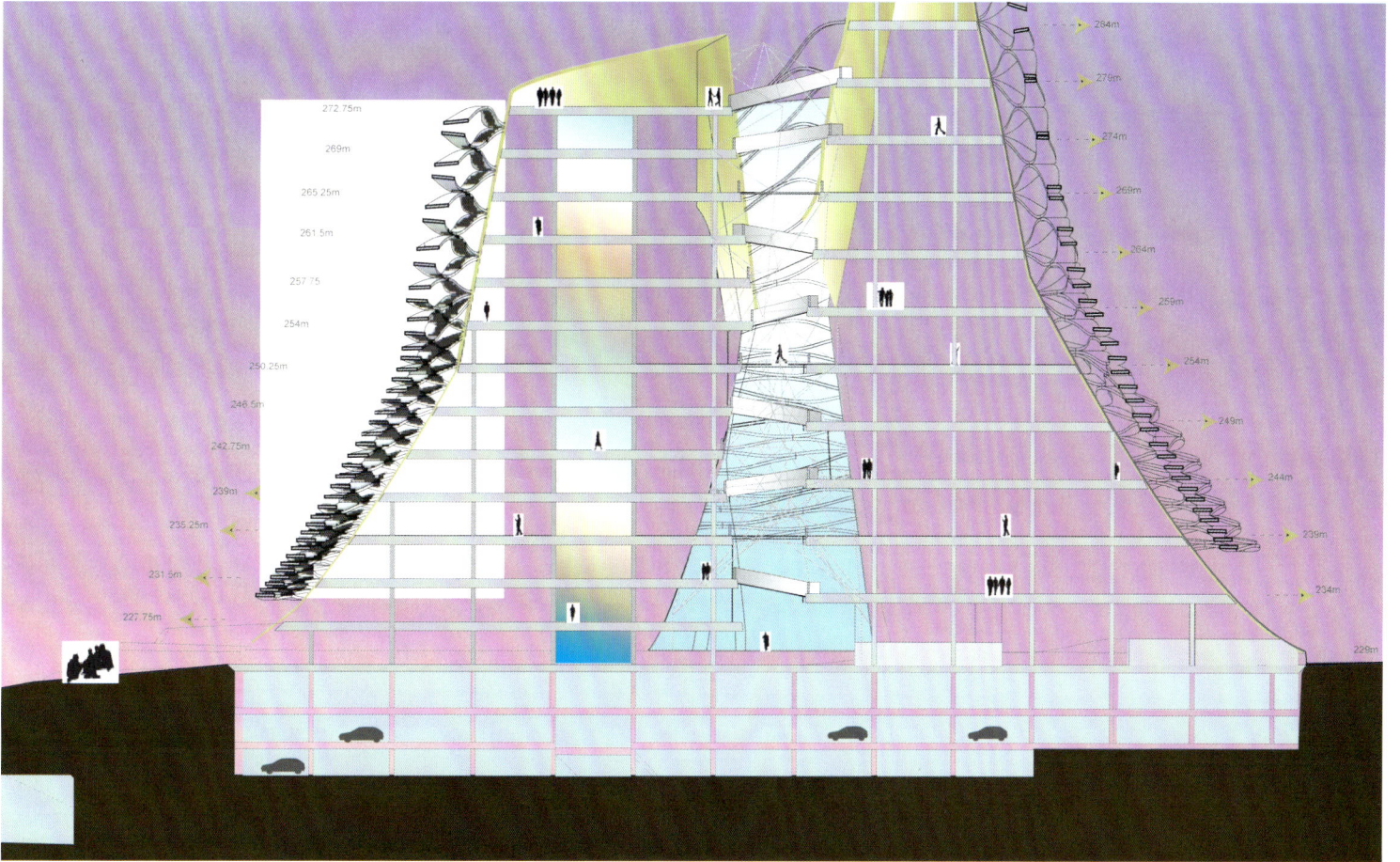

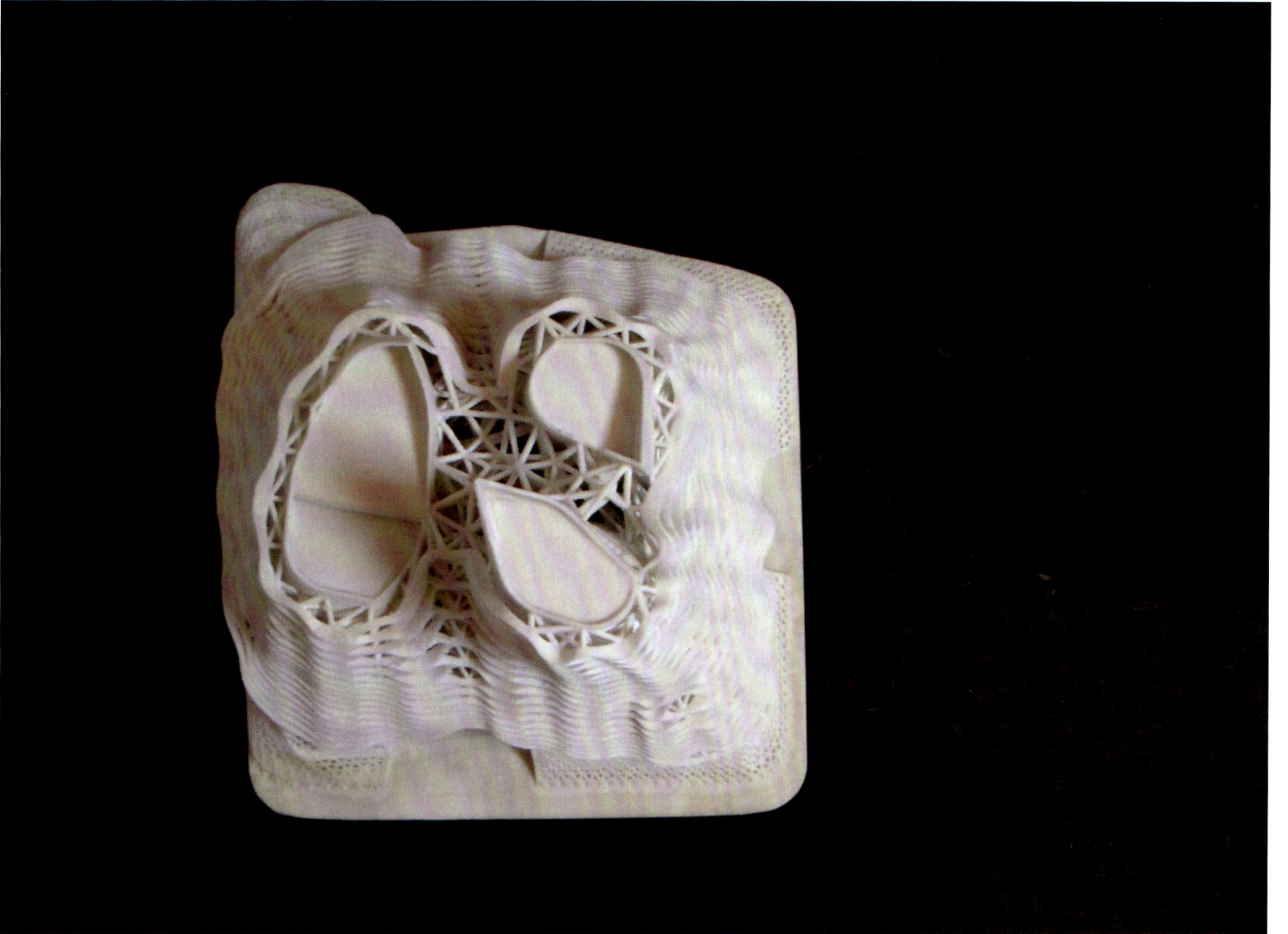

霍尔泽·科博建筑事务所
参赛项目8（第八名）

Holzer Kober Architekturen
Project 8 the eighth place

Atrium 中庭

Library hall 大厅

Reading rooms 阅览室

Kazakhstan Library Competition
哈萨克斯坦图书馆竞赛

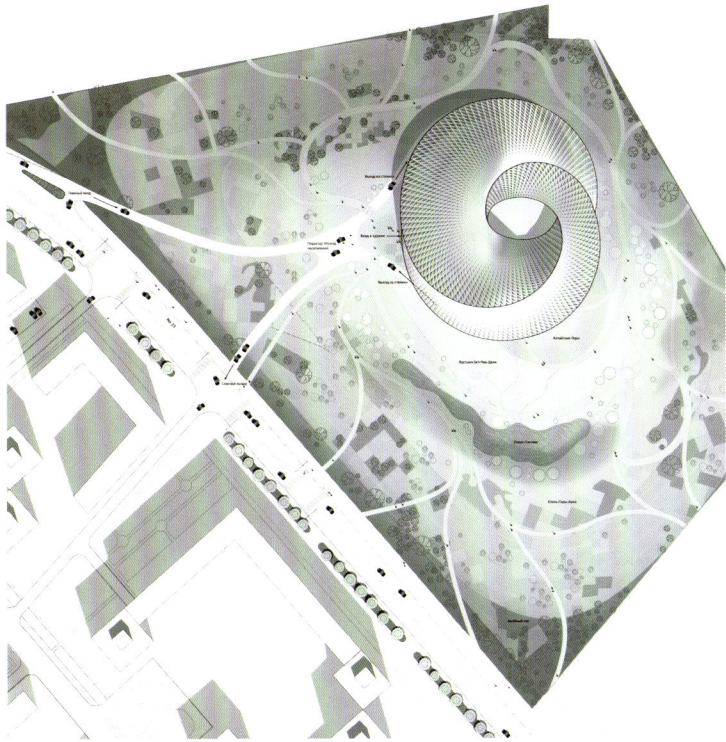

What public building better represents the budding aspirations of a new nation than a library? A part of the former Soviet Union and less than two decades old, Kazakhstan needed a strong cultural symbol to balance its reputation as just another oil-rich province in Central Asia. Kazakhstan as a nation should be taken seriously on the world stage.

还有什么建筑形式比图书馆更能表现一个国家崛起的渴望？作为前苏联的一部分，只拥有不到20年历史，哈萨克斯坦需要一个强大的文化象征来平衡它中亚地区石油大国的身份。在世界的舞台上，哈萨克斯坦不容被忽视。

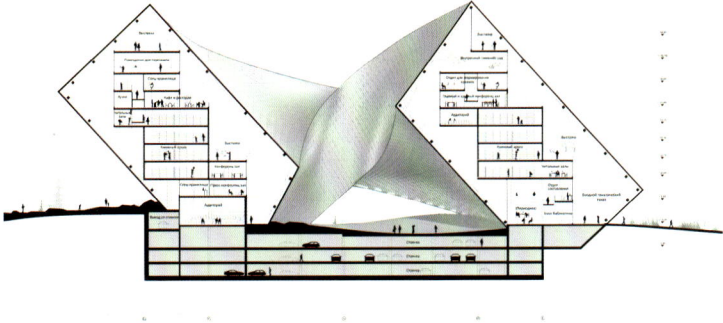

专家点评 **Remarks**

Of course, there were other designs which also featured circular schemes; but it would appear that the BIG entry's solution was more compact and and better organized than the other competitors. Kazakhstan will have a new library in the very near future.

当然，也有其他设计运用了环形结构，但是BIG的设计更为紧密、组织更为完善。哈萨克斯坦在不久的将来就要拥有一座新的图书馆了。

BIG 参赛项目1（一等奖） **BIG Project 1 the first prize**

BIG's winning entry bore a marked resemblance to their winning design for the Shanghai 2010 Expo. Although much larger in scale, the library also revealed a spiral as the primary circulation feature. As for form, it is conceived as a circular 'yurt,' a version of the original Mongolian tent-like structure.

Circulation pattern 流通样式

BIG的优胜作品和他们2010上海世博会的获奖设计有明显的相似性。尽管规模更大，图书馆也采用了螺旋结构作为主要流通特征。外观上，它则像一个蒙古包。

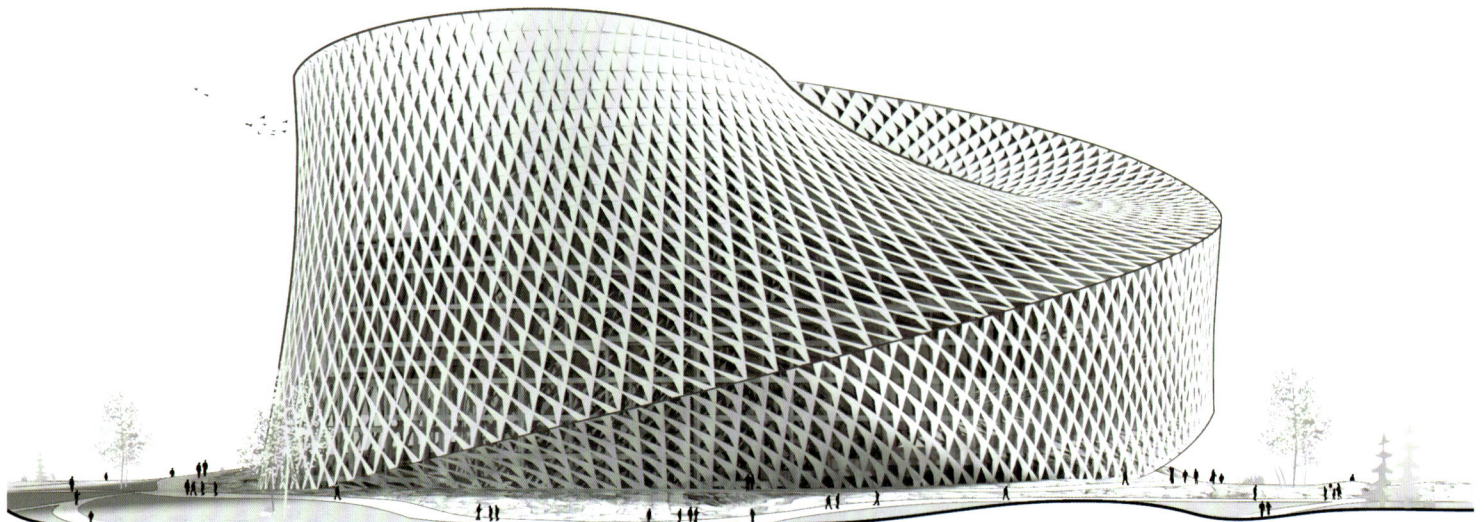

Elevation 立面图

View to entrance 入口处

Milford Library Competition

米尔福德图书馆竞赛

In 2004, Pike County Public Library (PCPL) received a donation from Dorothy E. Warner of the Milford estate for a new central library structure. Using this money, PCPL purchased a site in town that is surrounded by historic building and borders the Sawkill Creek. PCPL then formed a Building Task Force charged with selecting an architect and raising funds for the future state-of-the art library.

2004年，派克县公共图书馆从米尔福德的桃乐丝·E·华纳产业得到了一笔捐款，用于在米尔福德兴建一座新的中心图书馆。派克县公共图书馆用这笔钱购买了一块地皮，四周环绕着历史建筑和肖齐尔小溪的边界。随后他们组织了一个建筑特别工作组来挑选建筑师和为图书馆未来的建设筹集资金。

专家点评　**Remarks**

The jury's unanimously chosen scheme was designed by Frederic Schwartz Architects. "What sold everybody on Fred Schwartz's design was the fabulous atrium, the interior space he created. We felt he made the best use of the site, situating the library in a public green," states Maylene Syracuse, Chair for the Building Task Force. Alistair Gordon, jury member, says, "Their proposal really stood out of the pack; they had really thought through their proposal from soup to nuts. The interior was exactly what we wanted." Jujor Jonathon Marvel agrees, stating that, "the winning project best fit the site conditions, relating to the nature around it while offering a major interior public space."

评委们一致选择弗莱德里克·施华兹建筑事务所的作品为一等奖。"弗莱德里克·施华兹设计的卖点在于那出色的中庭和绝佳的室内空间。他将图书馆建在公共绿地上，我们认为最好的利用了场地。"建筑特别工作组组长梅里恩·赛拉克丝说。评委阿里斯代尔·戈登称："他们的设计在参赛作品中脱颖而出；他们的作品经过了从头到尾深刻的思考。室内设计正合我们的心意。"评委强纳森·马维尔说："获奖的项目极佳的契合了场地的环境，既和自然联系紧密，又提供了一个很好的室内公共空间。"

弗莱德里克·施华兹建筑事务所
参赛项目1（一等奖）

Frederc Schwartz Architects
Project 1 the first prize

The winning two-story proposal positions the building towards the back of the site, taking full advantage of the spectacular views across the ravine to the river. Parking has been pushed to the outer perimeter of the entrance allowing for a short allee leading to the library. The second story of the library is dramatically cantilevered, ten feet out in the front and 15 feet out in the back. This allowed the architects to bury the structure within the building with the walls acting as trusses. The focal point of the building's interior is an open sky-lit "hub" that will allow for large community meetings. A model of sustainability , the project proposes to follow LEED design principals.

"Rarely do you get a site that sits so perfectly on the American grid, " says Fred Schwartz, "We took unique advantage of the site by inserting our two-story building into nature. The front of the site is a public plaza, a public green." Inspired by the incredible town spirit and local public interest in the project, Schwartz likes to think that the future erection of the library will continue the Pennsylvania tradition of communal building.

Plan 平面图

获奖的设计将两层的建筑安排在场地的后部，从图书馆可以充分享受山谷和河流之间壮丽的景色。停车场被推到了入口的外圈，有一条小径通往图书馆。图书馆的二楼采取悬臂结构，前面突出3米，后面突出4.5米。这让建筑师把建筑结构藏在大楼里，用墙壁作为支撑的桁架。图书馆室内的焦点是开放自然采光的中心，那里可以举办大型集会。作为可持续性的典范，这个项目将以绿色建筑标准（LEED）打造。

"你很少会在美国找到一块这么好的场地"，弗莱德里克·施华兹说，"我们利用场地独特的优势将我们的双层建筑融入自然。场地的前部是一个公共广场，公共绿地。"被不可思议的城镇精神和当地公众对项目的兴趣所激励，施华兹认为图书馆未来的建设会延续宾夕法尼亚州公共建筑的传统。

View from ravine 从山谷看图书馆

PIKE COUNTY PUBLIC LIBRARY

View to entrance 入口处

PIKE COUNTY PUBLIC LIBRARY

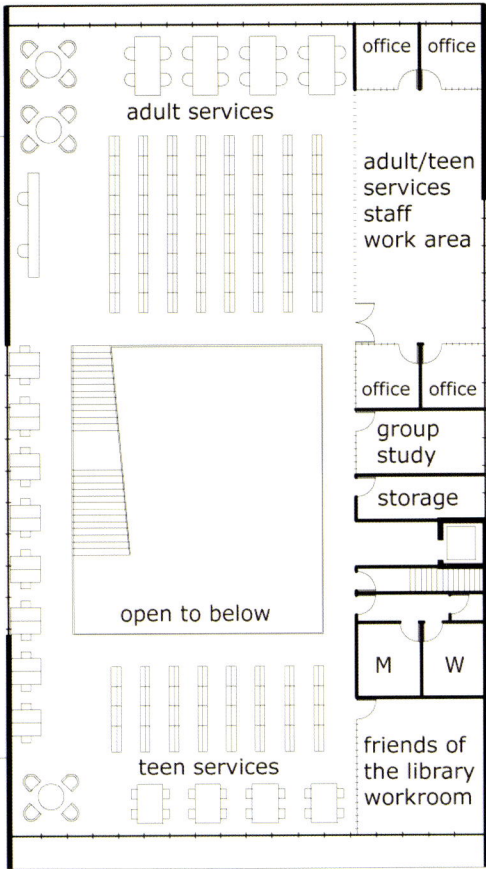

adult services

office office

adult/teen services staff work area

office office

group study

storage

open to below

M W

teen services

friends of the library workroom

PIKE COUNTY PUBLIC LIBRARY

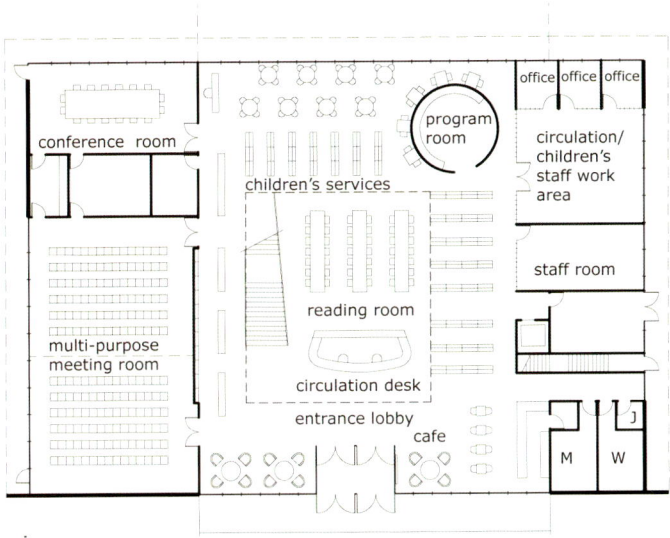

conference room

children's services

program room

office office office

circulation/ children's staff work area

staff room

reading room

multi-purpose meeting room

circulation desk

entrance lobby

cafe

M   W   J

博林·赛文斯基·杰克逊
参赛项目2（入围奖）

Bohlin Cywinski Jackson
Project 2 finalist

Bohlin Cywinski Jackson is best known as the designer of the trendy Apple Computer stores around the globe. Their entry in the PCPL competition, a low-key, modern duilding, offered a solution that accommodates the entire program for the library in a one-story structure that stretches across the majority of the site. Principal architect Peter Bohlin says, "We did a one-story scheme because it is the least costly to build and we felt this was the most workable solution from the library's point of view. A multi-story library doubles operating costs." This one-level plan does elimate the need for elevator, exit stairs and a substantial second floor structure while providing clear organization for supervision and legibility of the library. The community rooms are designed at a domestic scale, apparently not read as a plus by the jury, who were looking for larger scale communal spaces. As in the winning scheme, sustainablity issues were at the forefront of this design.

View from Harford Street 从哈尔福德街看图书馆

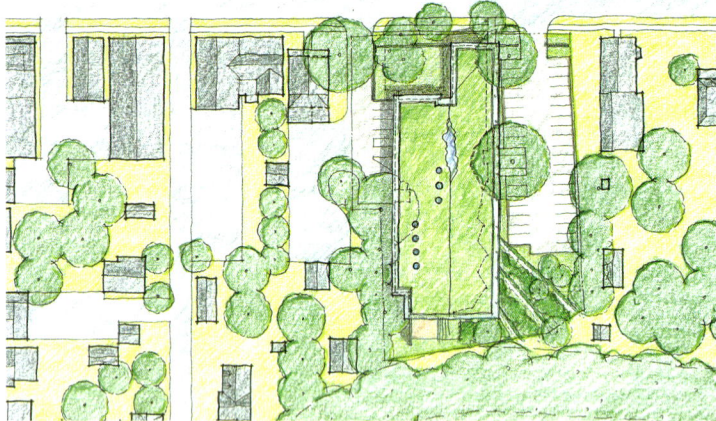

Plan 平面图

Section 剖面图

博林·赛文斯基·杰克逊公司以设计遍布全球的苹果电脑专卖店而闻名。他们参与米尔福德图书馆竞赛的作品是一栋低调的现代建筑。他们把图书馆的全套设施安放在一座横跨大半场地的单层建筑里。主建筑师彼得·博林说："我们之所以选择单层结构，是因为它的花费最低，同时以图书馆的角度来看，这是最可行的方案。多层图书馆会让建设成本翻一番。"单层的设计免除了电梯、楼梯和二楼的机构，而且为图书馆提供了清晰的监控系统和很高的辨识度。社区活动室被设计成了家庭规模，这显然不受评委的欣赏，因为他们需要更大的公共空间。和获胜设计一样，可持续性也是这个设计首要考虑的问题。

View from ravine 从山谷看图书馆

BKSK建筑事务所 参赛项目3（入围奖）

BKSK Project 3 finalist

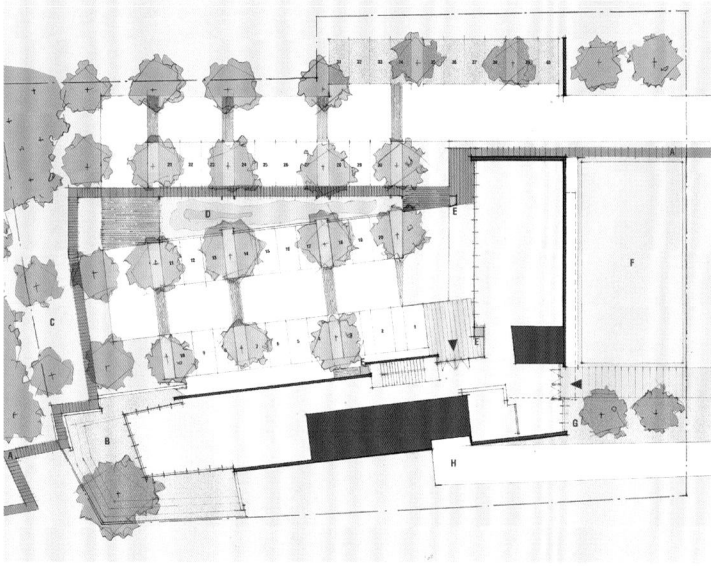

Plan 平面图

BKSK envisioned their three-story library proposal as a filter between the manmade town of Milford and the Sawkill River. The building footprint is defined by a classic L-shaped plan fronts the street while opening to the parking lot and storm-water retention pond in the rear. Clapboard siding and metal cladding as well as the central cupola structure were chosen to, "symbolize Milford as the birthplace of the American Conservation Movement," says Harry Kendall, principal at BKSK. In an unusual move, Kendall and his colleagues have incorporated an orchard and a bio-swale into the rear parking area, in order to, "break up the expanse of impermeable parking" Though the jury felt it was a conceptually compelling idea, they found it hard to imagine it as successful in reality.

BKSK设计的三层高的图书馆成为了人造的米尔福德城和天然的肖齐尔河之间的过滤器。建筑呈经典的L形，前端面向街道，后面是开阔的停车场和泄洪花园。设计采取了护墙楔形壁板和金属贴面，以及中心穹顶结构。哈里·坎达尔称："这象征着米尔福德是美国自然保护运动的发源地。"出人意料的，坎达尔和他的同事们在后部的停车区域设计了一个果园和生态湿地，目的是"用多样的步行地表打破不能通过的停车场的扩张"。这种想法很诱人，但是现实上很难成功。

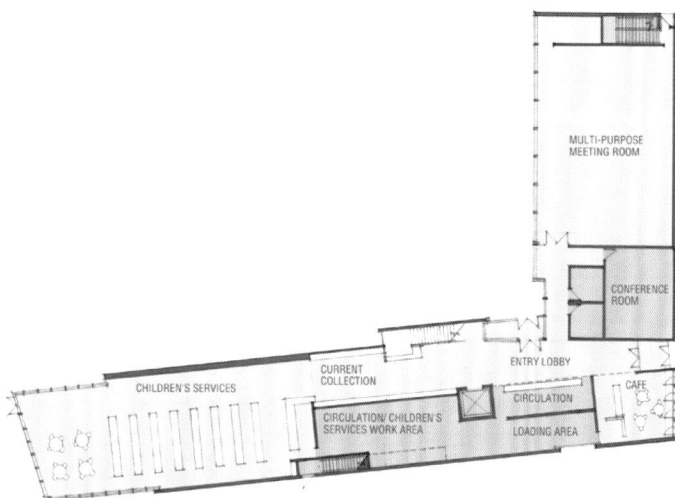

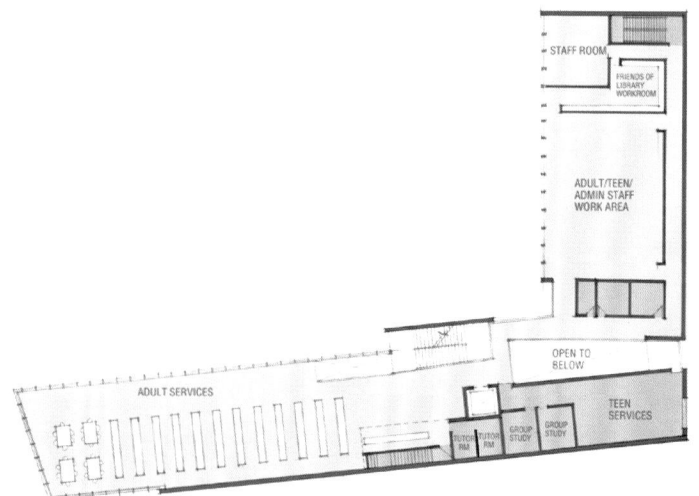

Elevation 立面图

Elevation 立面图

Section 剖面图

Philadelphia's Public Library Competition

费城公共图书馆竞赛

竞赛背景　Competition Background

The new breed of library requires a new building type that is less about quiet storage and more about dynamic interaction. Recently a competition was held to renovate and expand the Free Library of Philadelphia. At 180,000 square feet, the new addition is hardly a simple one and the design challenge was twofold: renovate the existing 75 year-old building and create a seamless (and presumably very rich) connection to the new facility. The Free Library of Philadelphia Foundation – a private entity assembled to manage the entire renovation project and they have chosen the invited competition format to select the architect.

新型图书馆的建筑需要一种新类型，其储藏功能要少一些、而动态互动要多一些。最近，费城自由图书馆举办了一次翻修和扩大建筑项目的竞赛。新建大楼的面积有1.68万平方米，建筑起来绝非易事，设计的挑战有二：一是对有着75年历史的旧楼进行翻修，二是使旧楼与新楼实现无缝连接。为此成立了费城自由图书馆基金会，由这个私人性实体来管理整个翻新项目的运作，为了找到最好的建筑事务所，他们已经选择了邀请建筑师事务所参加的竞赛程序。

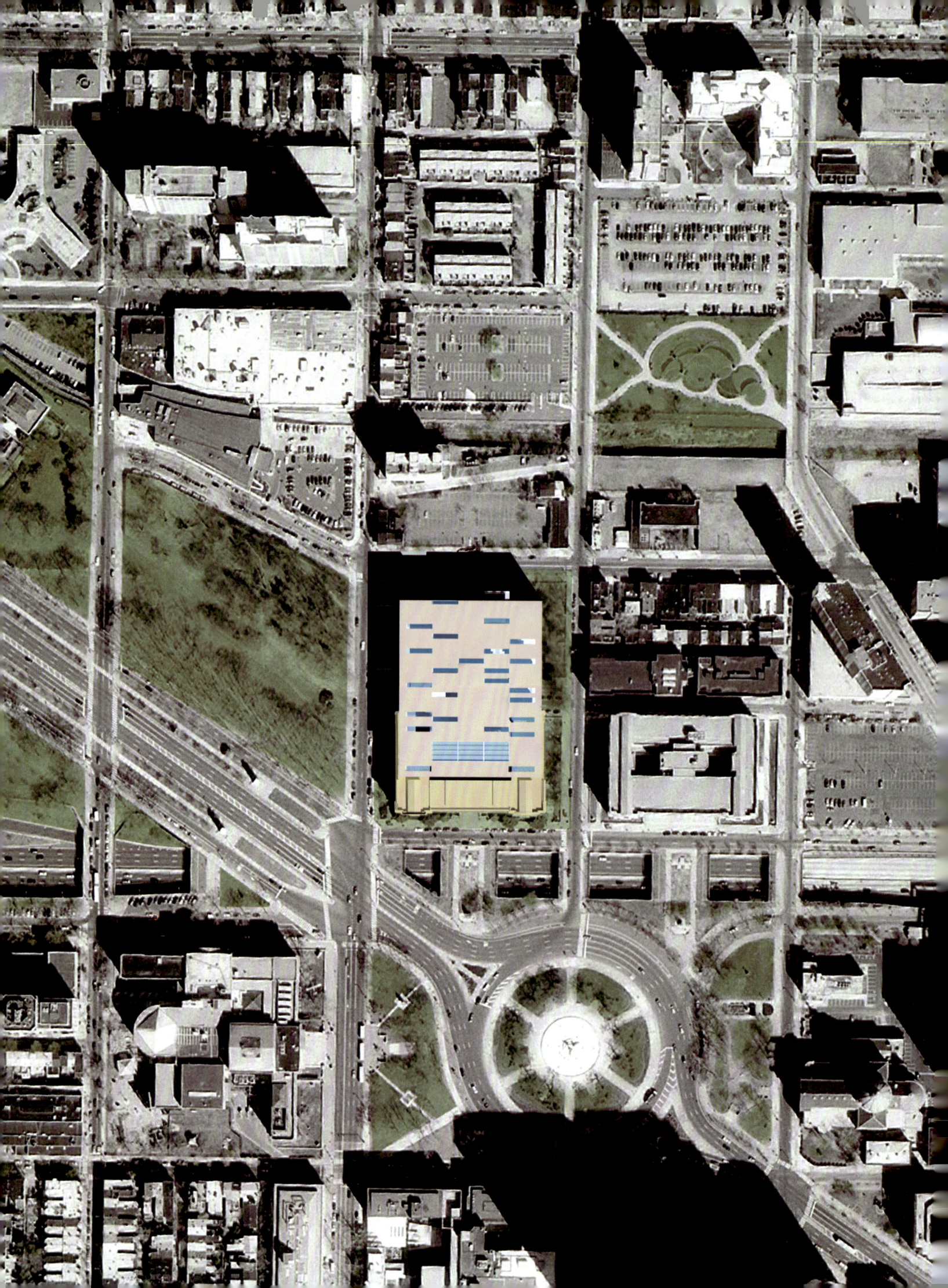

专家点评 **Remarks**

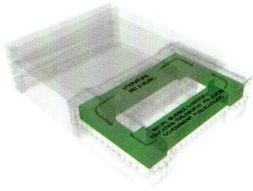

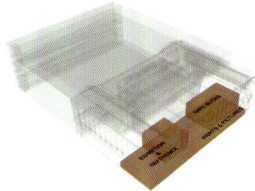

The finalists all focused on very different issues, addressing everything from urban design strategies to modern social phenomena, but each produced a programmatically successful and spatially rich building. Consciously or not, they each made a statement about the existing library building, as well. Size, shape and materiality all effect the general public's perception of the old and the new – as both a structure and an institution. It may be then, that Safdie's slightly less aggressive concept has a built-in reverence for the old building, which in turn could be read as a better understanding of the history and purpose of the American library. Moshe Safdie's new Philadelphia Library will communicate a sensitivity and respect for the history and purpose of one of our greatest institutions.

几家入围建筑事务所注意到了不同的问题，都从城市设计出发解释了现代社会现象，每一个方案都产生出一个规划成功、空间内涵丰富的建筑，每一个都对现有图书馆大楼作出了陈述。其规模、形状和材料影响了公众对于既是建筑又是办公机构的新老图书馆的看法。萨菲德的概念突出表现出对老建筑的崇敬，因而被看成是对历史和美国图书馆用途的更好的诠释。萨菲德建筑事务所的新费城图书馆项目，传达出一种对历史的敏感与尊重和一个最好的公共机构的意图。

摩西·萨菲德建筑事务所
参赛项目1（一等奖）

Moshe Safdie & Associates
Project 1 the first prize

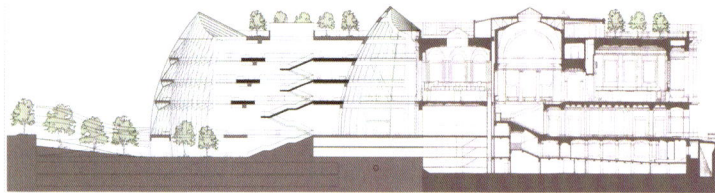

Section 剖面图

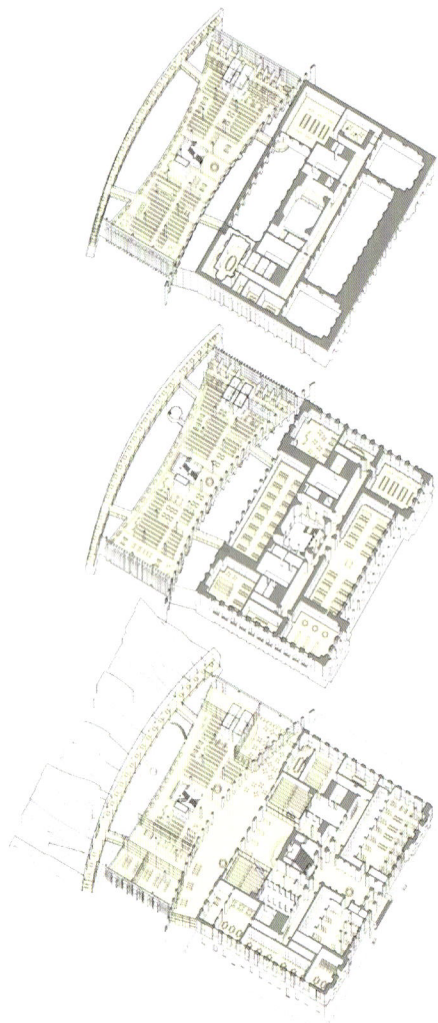

Safdie relies on a flood of daylight to animate the interior spaces, while also playing with some structural ambiguity, which all but erases the line between inside and out. He proposed a central atrium – which he refers to as the "urban room" – as the link between the existing and the new structures. The anchor-piece of this space is the curving wall that torques on its way through the central axis, forming the roof of the atrium as well as creating the southern edge of the new book hall. This wall component runs beyond the planes of the east and west facades, partially embracing small exterior spaces. The main volume of the addition is sandwiched between the central atrium and the north-façade atrium and these two atria pinch in the core of the book hall, giving it a planometric waistline. This allows daylight deeper into the space and also gives the building its transparent quality. The north atrium is now the entrance to the building, as well as the "urban room", while the central atrium that connects the old and the new spaces has become the main reading room for the popular library. This is more effective. The north atrium will house shops and cafes, an auditorium, meeting rooms and several other, more social spaces. In the renderings produced for the competition, the interior is a soaring glass and steel space. Bridges link the seating areas on the outside wall to the main book hall on the interior and the ground floor opens up to a garden space on the north side. Similarly, in the central atrium, lightweight bridges provide upper story access between the new book hall and the classical façade of the original building.

摩西·萨菲德建筑事务所的建筑方案强调充分的自然采光，赋予室内空间以活力，同时也运用一些结构的模糊性，消除内外的界限。他提出用一个中庭连接外部环境与新建大楼，他本人把这个中庭叫做"城市空间"。这个空间的固定部分是扭转通过中轴的一个曲线墙，形成中庭的屋顶，也构成新建的读书大厅的南部边缘。这面墙位于东西两面建筑立面的平面之上，把一些小的外部空间全部收纳进来。新楼的主体夹在整个建筑的中庭和北面的建筑立面的中庭之间，这两个中庭夹在读书大厅的中心，形成平面的墙体围线。这样，日光照到空间里的面积就更大一些，从而增加建筑的通透感。北部中庭是整个大楼及"城市空间"的入口，而连接新旧两楼的中庭成了通俗图书馆的阅览室，这样安排更有效率。北部中庭将设置一些商店和咖啡馆、一个礼堂、一些会议室、几个其他用途的空间及更多的社交空间。向竞赛提交的效果图中，内部是一个高耸的玻璃钢架结构，连桥把外墙的座位区与内部的主体读书大厅连接起来，一楼向上直通到北部的一个花园。同样，在中庭，轻质的连桥在新建读书大厅和原有建筑的古典立面之间建立连接，人们可以上到上层，主楼层成为两个建筑共享的外廊。

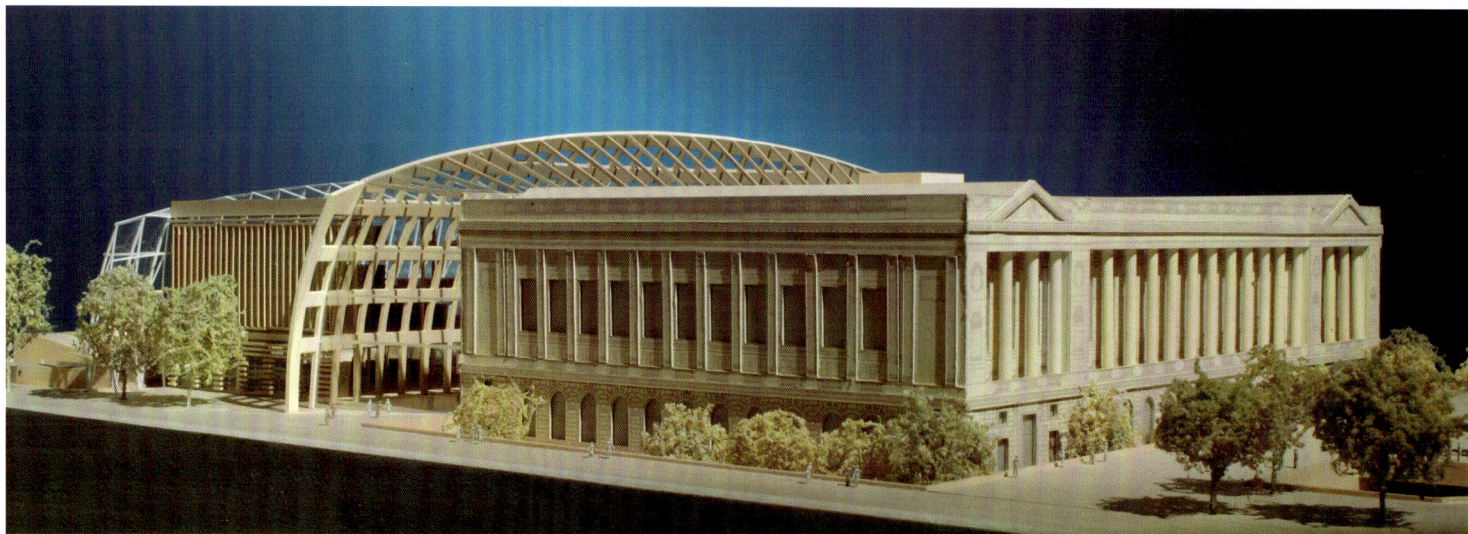

Perspective of atrium 中庭

View from street 从街上看图书馆

Atrium 中庭

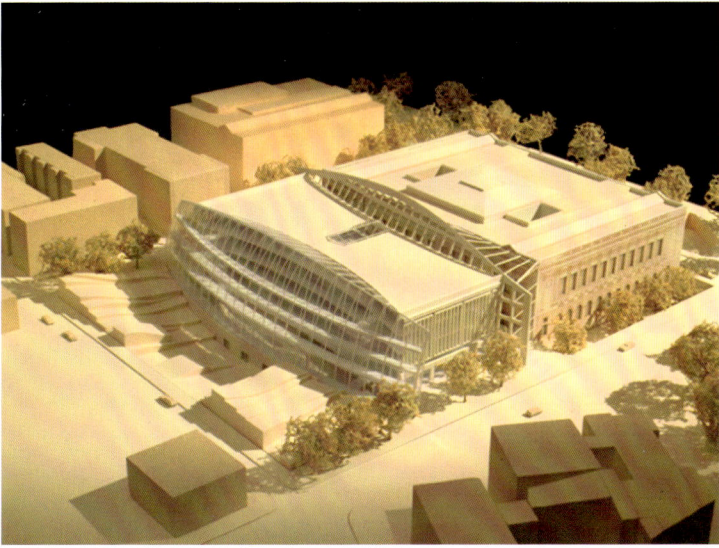

Evening view of model 模型夜间图

十人建筑事务所 参赛项目2（入围奖） **TEN Arquitectos Project 2 finalist**

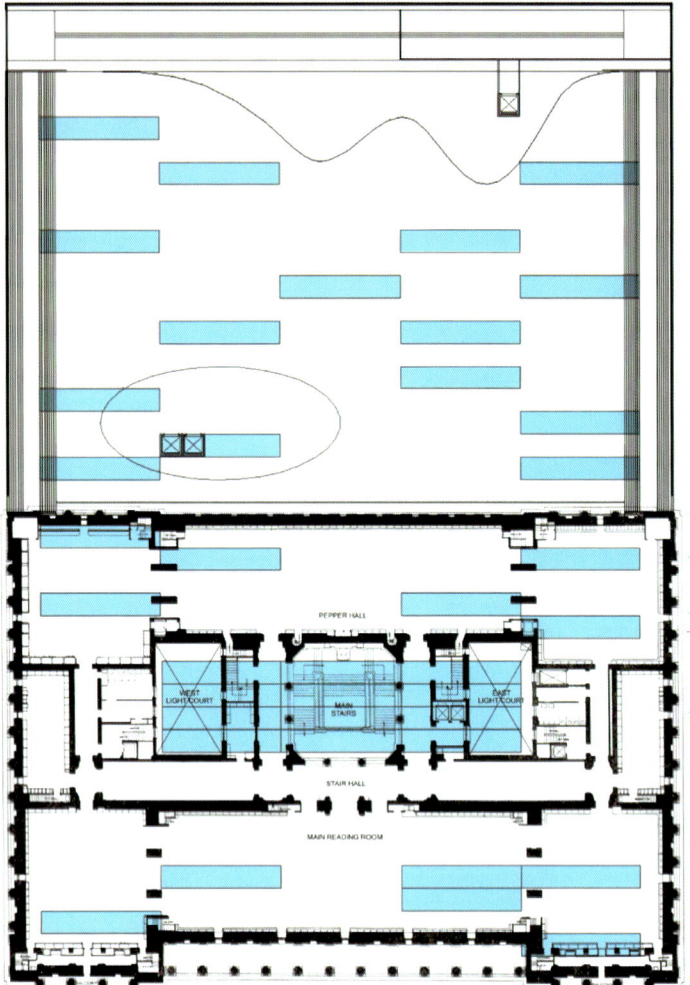

Section 剖面图

The proposal submitted by Enrique Norten's office, TEN Arquitectos is essentially a large inverted "L" shape that grows out of the northern edge of the site and crosses over to the existing library, fluidly joining the underground common areas to the vertical book storage to the techno-conference-bridge-space above, which Norten refers to as the "urban room"; however, he has chosen to leave the space open to the elements. This move allows the residual end of Fairmont Park to pass through the building, uninterrupted, creating a relationship between inside and outside that makes the existing, 'preserved' ground plane read as somewhat precious. The central plaza space is actually a new ground: a slab that is "perforated" at various points throughout, which provides access and light to the below-grade portions of the library. The inside face of the "L" at the northern end of the plaza is designed as a curving glass curtain that provides views into both the upper levels that house the books, as well as the common areas below the main plaza slab. The new building is organized so that the high volume stacks start in the sub-grade level and continue up the vertical bar to the elbow at the upper floor. The southern most edge of the bridge is reserved for a conference and business center and its location gives it a somewhat private feel. This conference area re-activates an old roof terrace on the existing building and completes the "L" shape with a dramatic viewing platform. Like the slab of the plaza below, the roof of the bridge component is also perforated with skylights, although quite a few more. Norten's design for Philadelphia relies on a large, dramatic building form that effectively solves the programmatic riddles on the inside, while the negative spaces created by the building's exterior become highly charged pieces of the urban fabric.

十人建筑事务所的埃里克·诺顿事务所的设计方案主要是在项目地点的北部边缘建起一个大大的倒置的"L"形结构，一直横跨到原来的图书馆大楼，把地下公共空间和地上的垂直书架以及技术处理室、会议室、连接桥连在了一起。诺顿也把这个"L"形之下的新空间称为"城市空间"，但是他选择把这个空间与其他部分连接起来。这种提议使费尔蒙特公园的残留部分不受任何阻碍地穿过大楼，在内部和外部之前建立联系，使得原有的"被保留下来的"地平面变得珍贵起来。中心广场空间实际上是一块新地：在一个厚板子的不同位置打上孔，使得光线能够进入图书馆下层的空间部分。广场北端的"L"形结构的内表面被设计成一个弯曲的玻璃幕墙，很容易就可以看到藏书的上层以及在主体广场厚板之下的公共区域。新建筑的构造便于高容量的书库从地基一层开始沿着垂直钢条一直排到上层楼面的弯管处。建筑的最南部边缘部分预留作会议和商务中心，它的所在位置给人一种私密感。这个会议区使原有建筑的一个屋顶平台重新发挥作用，用一个观景平台完成了"L"造型。像下面广场的厚板一样，在连桥屋顶也打孔安上了天窗。诺顿为费城图书馆所做设计的基础是一个大型的、引人注意的建筑结构，有效地解决内部程式化的难题，建筑外部的周围实体空间也成为充满动感的城市结构部分。

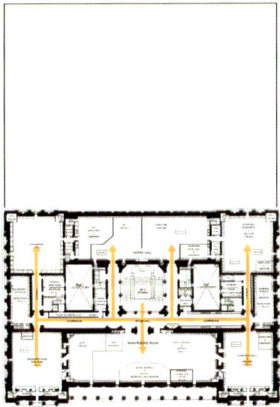
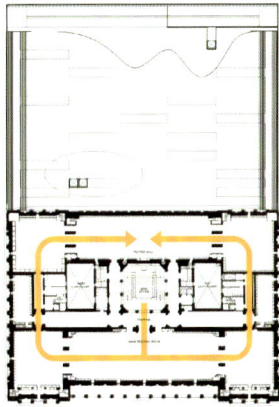
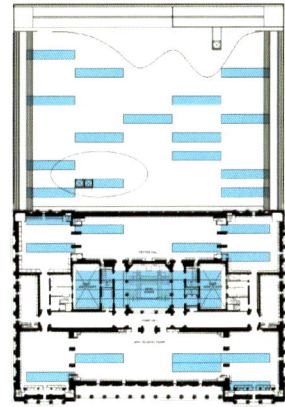

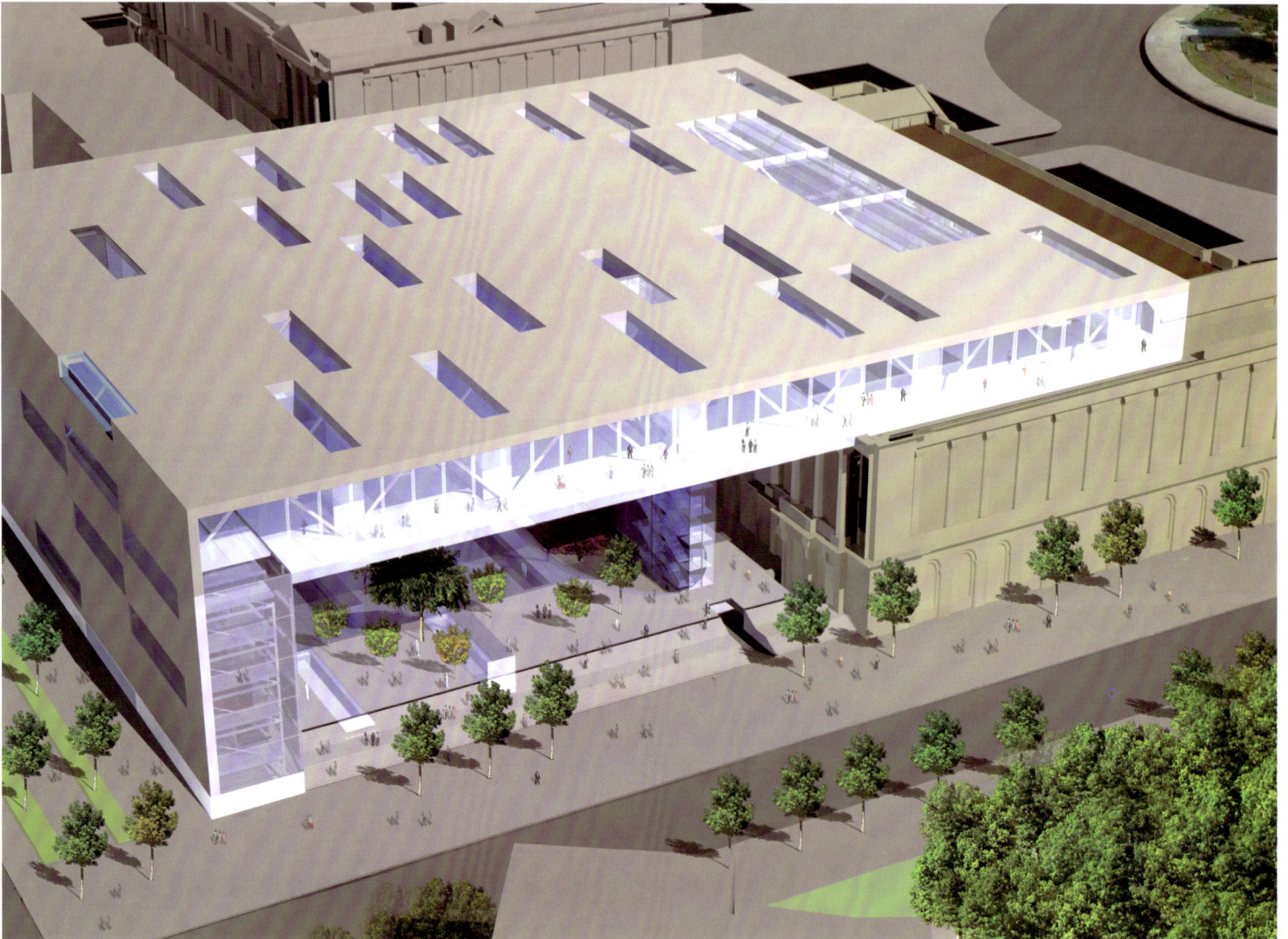

Aerial view 鸟瞰图

View to interior plaza 室内广场

Stockholm City Library Competition
斯德哥尔摩城市图书馆竞赛

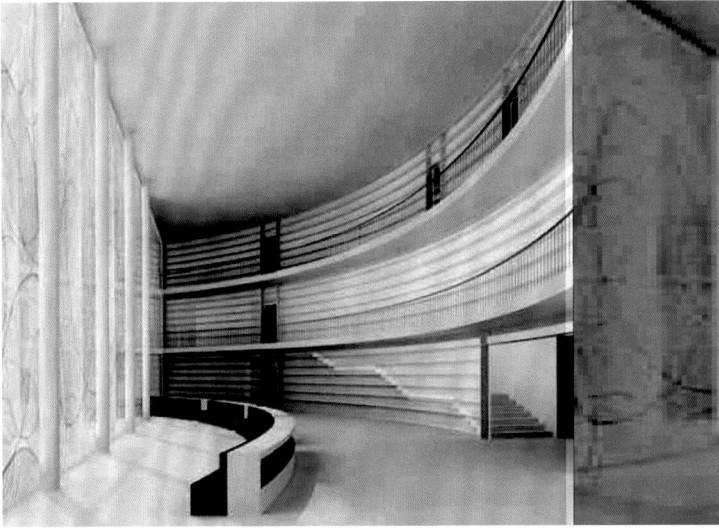

May of 2006, the City of Stockholm commissioned an open competition for an addition to the Stockholm City Library. This was conducted in two phases, and required a program of a total of 24,000sm — including the 14,000 sm from the original structure. The main goal, as outlined in the competition brief was to find a "high-class architectonic composition of which the library is an integral part."

Four criteria were decisive in determining the winning entry. First, the structure needed to be a distinguishable piece of architecture, while respecting the iconic library. In addition it had to be able to merge new technologies with existing cultural and historical values, as well as guarantee long-term adaptability. Finally, the new composition was required to complement the city's current and future urban fabric.

2006年5月，斯德哥尔摩市为斯德哥尔摩城市图书馆扩建发起了一次公开的建筑竞赛。竞赛分为两个阶段，设计总面积为2.4万平方米（包括原图书馆的1.4万平方米）。竞赛摘要中描述的主要目标是找出一个"以原图书馆为主体的高层次混合结构"。

评选优胜作品有四个条件：首先，结构既要有区别性，又要尊重原有建筑。其次，新技术的运用必须和现有的文化和历史价值相结合。第三，要保证建筑的持久性。最后，新的混合结构要补充到城市当前以及未来的城市结构之中。

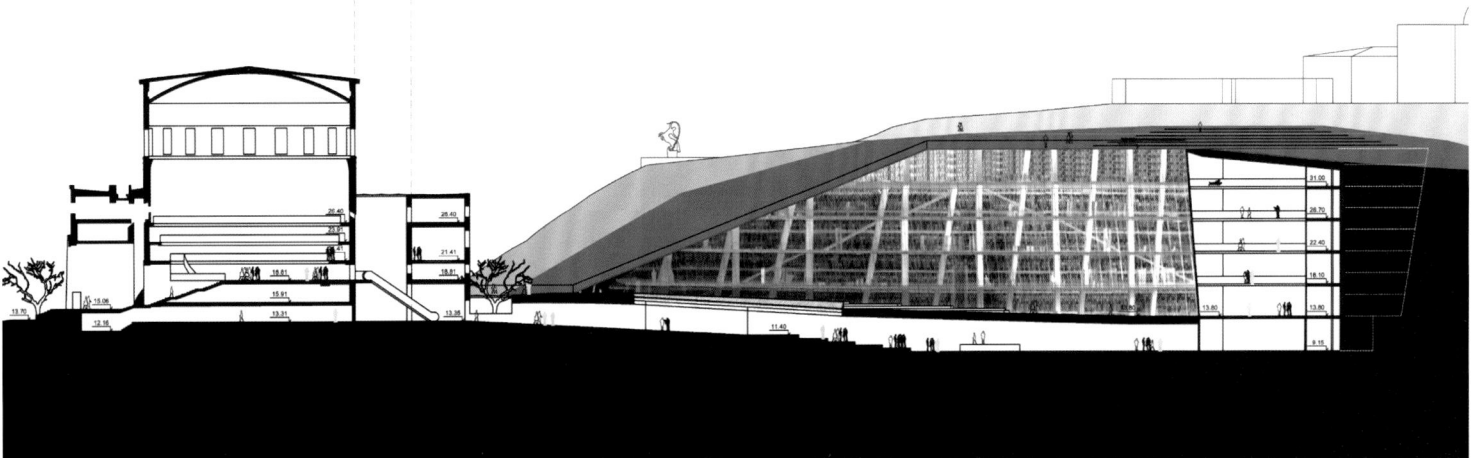

专家点评 Remarks

Hanada's design resembles Sweden's national monuments in that it merges cultural and building precedents to produce architecture with links to the past. It clearly and gracefully addresses all of the jury's main concerns, providing Stockholm with a respectful, yet bold new landmark.

哈纳达的设计和瑞典国家纪念馆一样，仿效文化建筑的先例创造连接过去的建筑。它明确而漂亮的完成了评审委员会的要求，为斯德哥尔摩提供了一个受人尊敬、又大胆创新的地标。

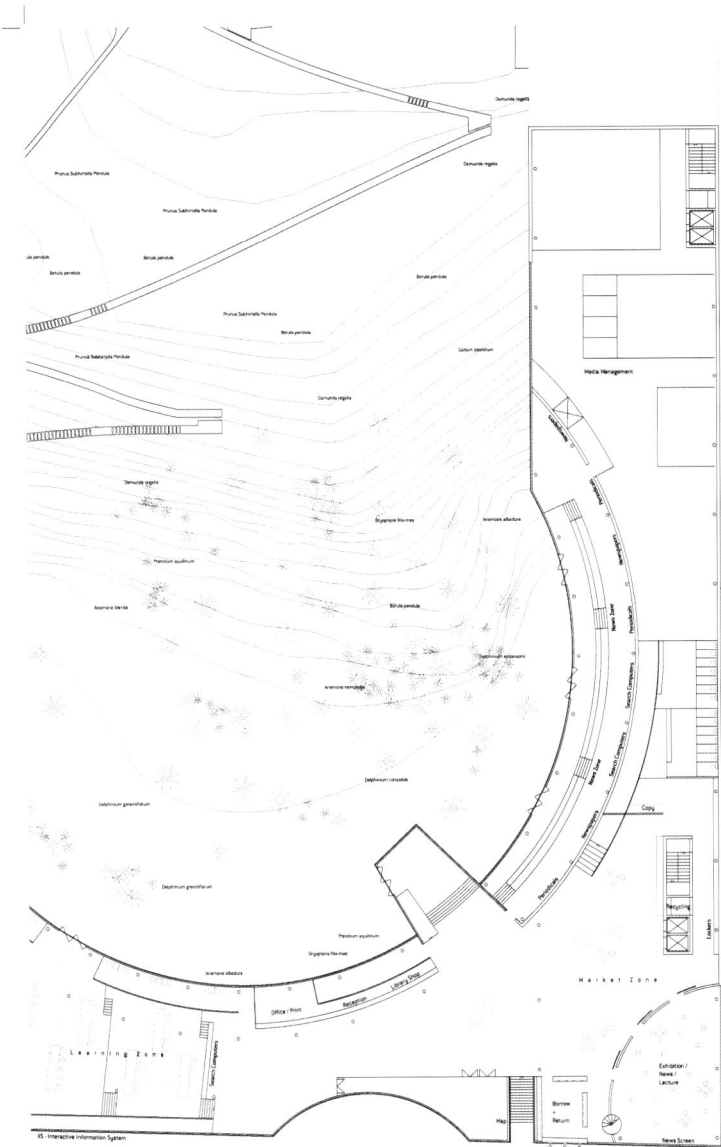

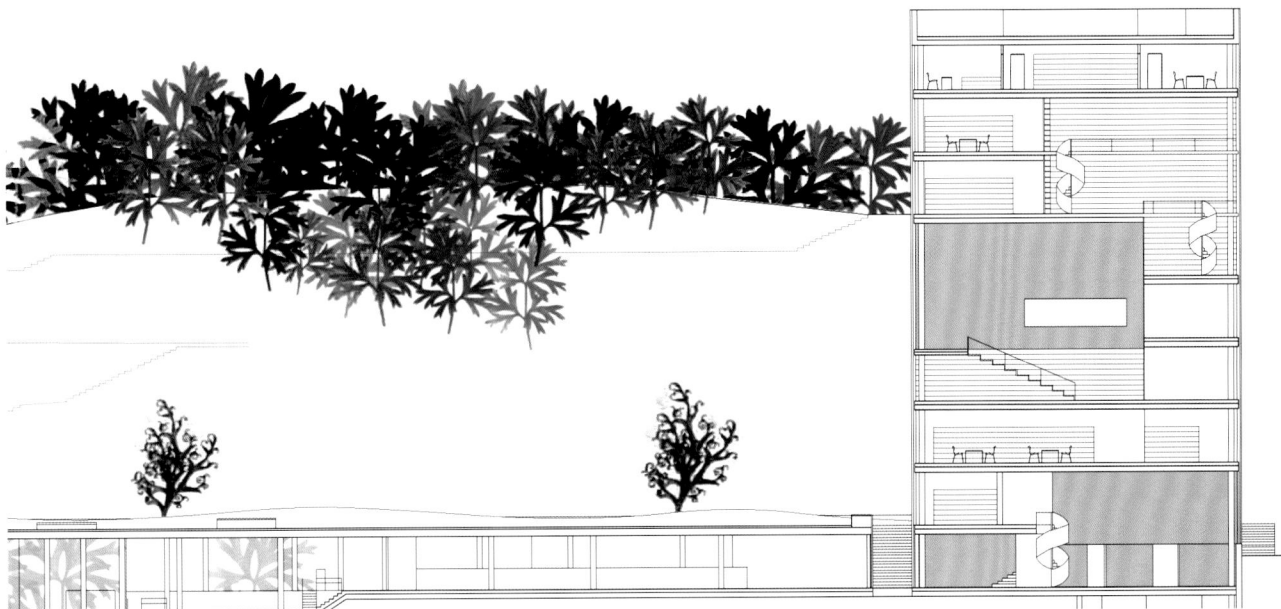

黑科·哈纳达 参赛项目1（一等奖）

Heike Hanada Project 1 the first prize

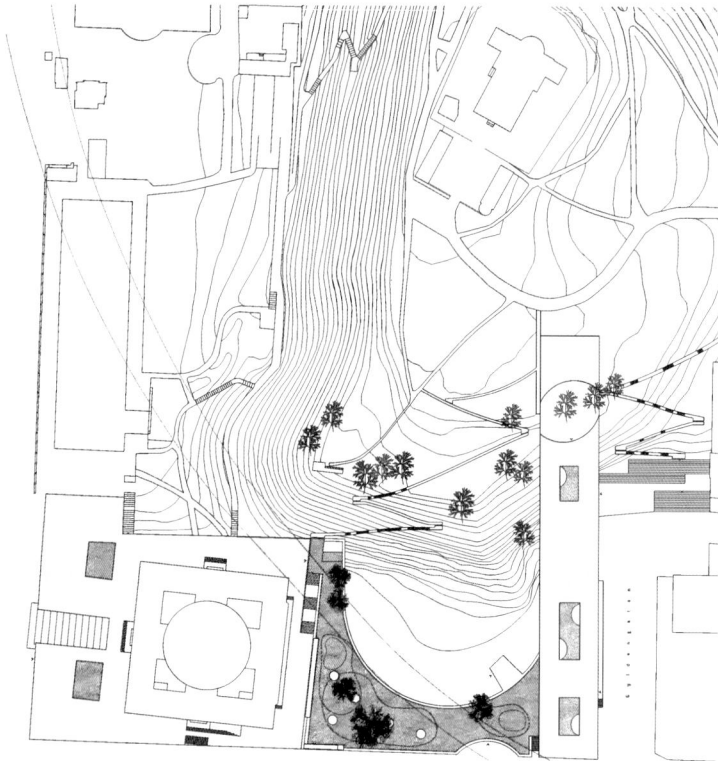

Site plan 总设计图

The design by German architect Heike Hanada best met the jurie requirements.Based on the drawing and a persuasive night perspective, it was clear that this work would be a distinct architectural additon to the city. Its luminous façade is in stark contrast with the heavy, brick structure of other public buildings. At first glance, it is difficult to find the correlaton between Asplund and Hanada; yet it exists in the simplicity and detail of the design. Like Asplund, Handa concentrates her work in regular geometric volumes. Heike Hanada subtly links the two structures without replicating the original work.

Most importantly, the proposal shows great respect for the existing structure. The low entrance area allows for two separate buildings to emerge. The main structure is located along the least important part of the hill, and its uniform façade provides an elegant background for Asplund's Library. In terms of the obervatory and the hill, the project opens up to both with a semicircular garden. This communal zone serves not only to gather and redistribute people throughout the buildings, but also to attract passerby from the transit stations to its periodicals room for café. The transparency of the building's façade can de regulated to either expose all the interior activities or to simply reflect the hill and its foliage. Essentially this gives the building the magical power to appear or disappear within the city's fabric.

德国建筑师黑科·哈纳达的设计最符合评审委员会的要求。从设计图和一张夜间效果图来看，这个设计将为城市添加一个独特的建筑。它发光的外墙将和其他公共建筑沉重的砖结构形成鲜明对比。一开始，也许很难找出哈纳达设计和阿斯普兰德（原图书馆设计者）设计的相同点，其实它存在于设计的简单和细节中。和阿斯普兰德一样，哈纳达将工作集中在几何图形空间上。黑科·哈纳达没有复制原图书馆，而是巧妙的将两栋建筑联系了起来。

最重要的是，设计对原建筑显示出了极大的尊重。较低的入口区让两栋独立的建筑融合在一起。新建筑的主结构建在山上最不显眼的地方，成为了原图书馆优雅的背景。面向天文馆和小山，新图书馆开放了一个半圆形的花园。这个公共区域不仅聚集和重新分配了图书馆的人流，也吸引了公交站的路人到图书馆的期刊阅览室和咖啡厅。透明的建筑外墙是可调节的，既可以看到室内结构，又能倒映小山和植物。这让建筑有了魔力，可以随意消失在城市结构之中。

View to learning zone 学习区

Aerial perspective at night 夜间鸟瞰图

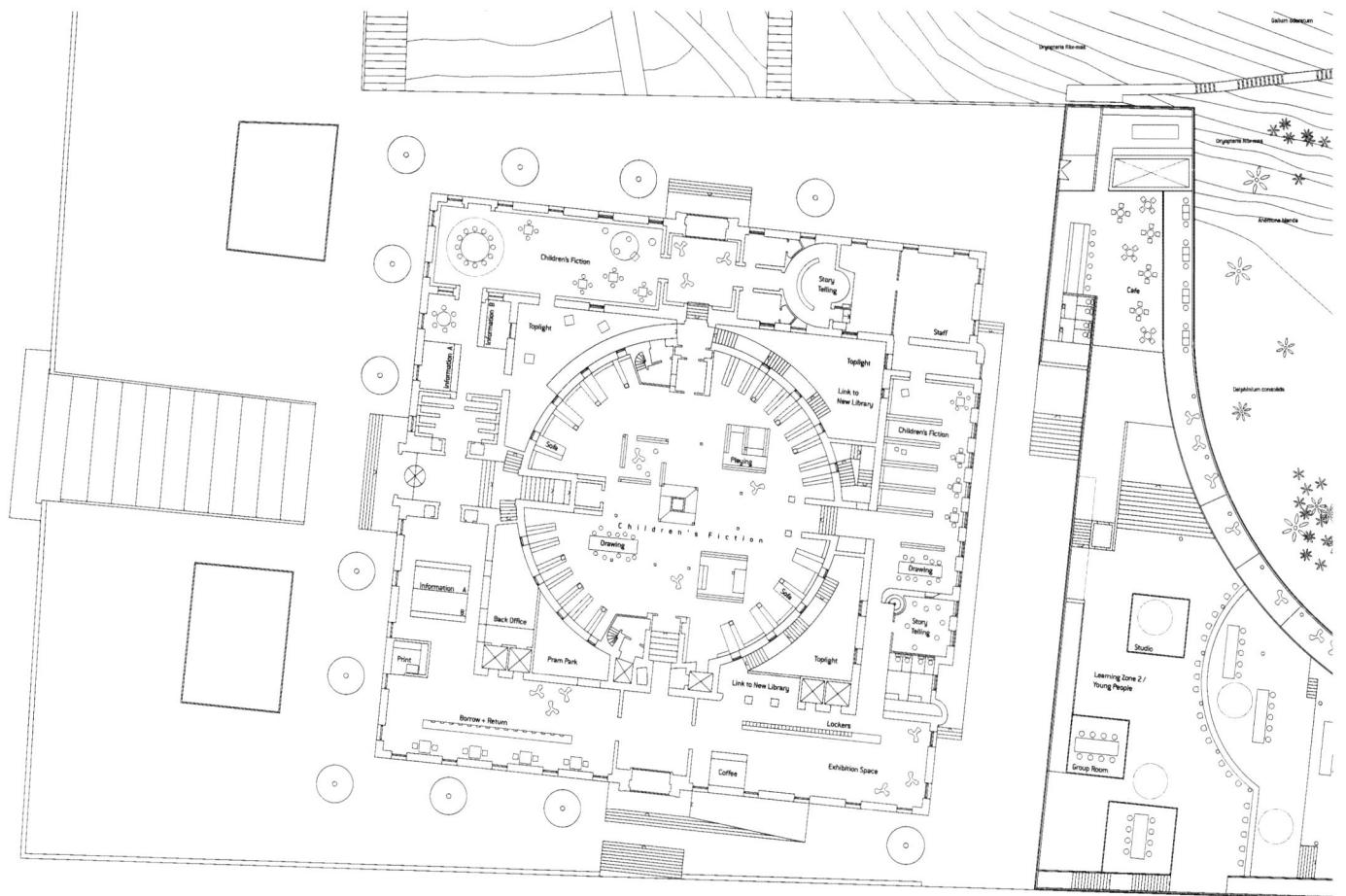

Floor plan 平面图

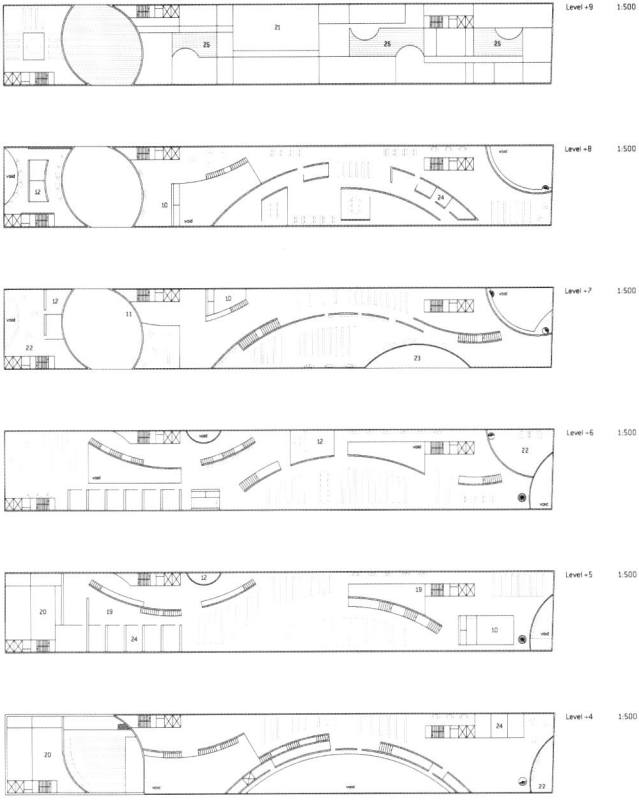

Level +9   1:500

Level +8   1:500

Level +7   1:500

Level +6   1:500

Level +5   1:500

Level +4   1:500

其他竞赛方案 **Other contestants' schemes**

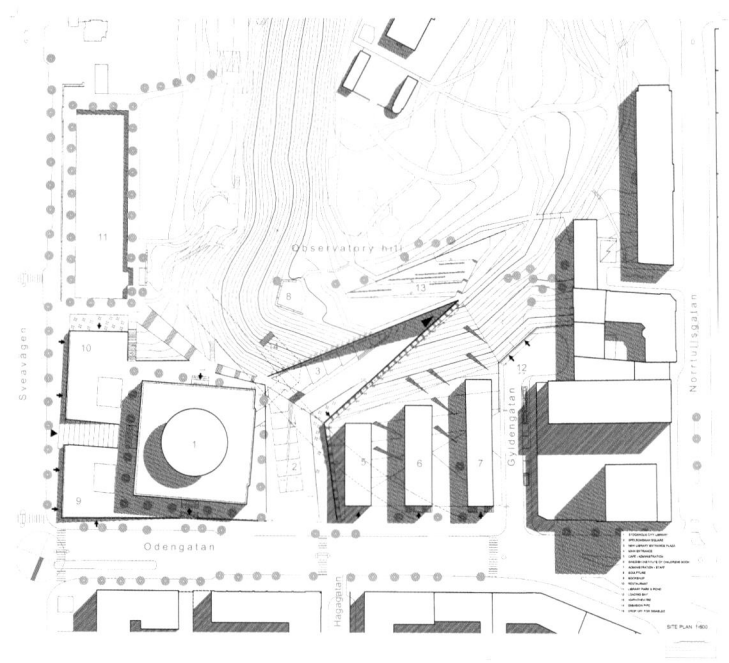

Site plan 总设计图

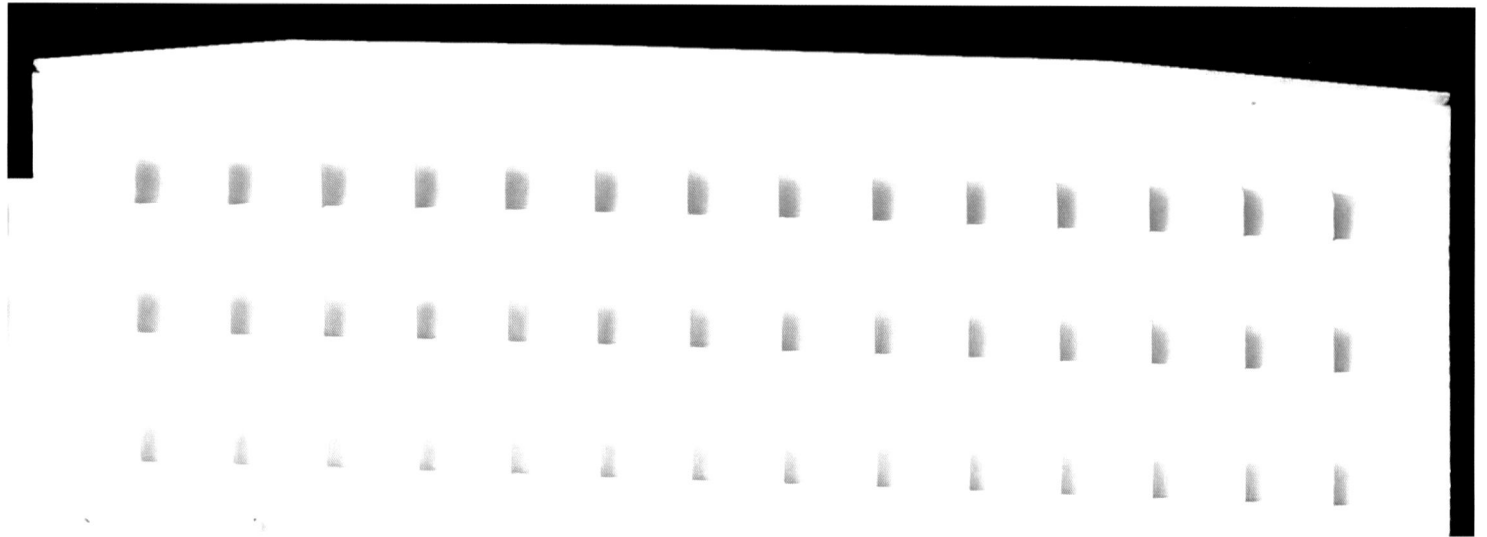

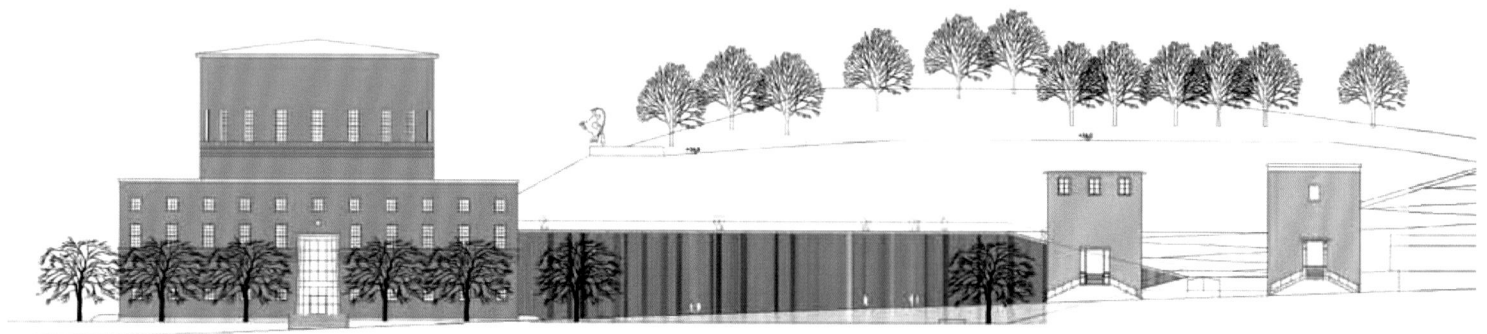

Elevation 立面图

Reading room 阅览室

INDEX

Darat King Abdullah Ⅱ Arts Center Competition

Zaha Hadid Architects
Tel: +44(0)20 7253 5147
Fax: +44(0)20 7251 8322
Delugan Meissl Associates Architects
Tel: +1.585 3690
Fax: +1.585 3690 – 11
Snøhetta AS
Tel: +47 24 15 60 60
Fax: +47 24 15 60 61
Kerry Hill Architects
Tel: +65 6323 5400
Fax: +65 6323 5411
Christian de Portzamparc
Tel: +331 40 64 80 00
Fax: +331 43 27 74 79
Henning Larsen Architects
Tel: +45 8233 3000
Fax: +45 8333 3099

Estonian National Museum Competition

Dorell.Ghotmeh.Tane / Architects
Tel: +33 1 43 38 12 47
Fax: +33 1 43 38 12 85
ALA Architects
Tel: +358 9 4259 7330
Fax: +358 9 4259 7331
Bramberger Architects
Tel: +43 316 269 377 10
Fax: +43 316 269 377 11
Local Architecture
Tel: +41 21 320 06 86
Fax: +41 21 320 06 86
Arkkitehtuuri oy lehtinen miettunen
Tel: 03 652 5270
Fax: 03 652 5200

The Gyeonggi-do Jeongok Prehistory Museum Competition

X-TU Architects
Tel: 00 33 1 45 23 37 10
Paul Preissner Architects
Tel: (312) 593-4177
Easton + Combs
Tel: 001.347.410.9088

Mammoth and Permafrost Museum Competition

Soren Robert Lund Arkitekter
Tel: +45 33910100
Fax: +45 33914510

Michigan State University Art Museum Competition

Zaha Hadid Architects
Tel: +44(0)20 7253 5147
Fax: +44(0)20 7251 8322
Coop Himmelb(l)au
Tel: +43 1 546 60 – 0
Fax: +43 1 546 60 – 600
Kohn Pedersen Fox
Tel: 44(0) 20 3119 5300
Fax: 44(0) 20 7497 1175
Morphosis
Tel: +1 (310) 453-2247
Randall Stout Architects
Tel: 310.827.6876
Fax: 310.827.6879

Museum for L'Universitiare catholique de Louvain Competition

Perkins + Will
Fuksas Architects
TECTONICS ARCHITECTS Ltd.
Charles Vandenhove et Associes
Tel: +32 4 222 37 37
Fax: +32 4 222 17 21

Nam June Paik Museum Competition

Kirsten Schemel Architekten BDA
Tel: +49 (0) 30 8872 4956
Fax: +49 (0) 30 8872 4958
Kyu Sung Woo Architects
Tel: 617 547 0128
Noriaki Okabe
Tel: 03-3555-0610
Fax: 03-3555-0611
Schneider + Sendelbach Architekten
Tel: +49(0)531-244090
Fax: +49(0)531-2440925
Deubzer König Architeckten
Tel: +49 30 885 22 01
Fax: +49 30 885 22 05

San Jose State University Art Gallery Competition

WW Architucture
Tel: (713) 521-0529
Fax: (815) 572-8249
SPF: architects
Tel: 310 558 0902
Fax: 310 558 0904
Eight Inc.
Tel: +1 415 434 8462
Swanke Hayden Connell Intl. Ltd.
Tel: +1 212 226 9696
Fax: +1 212 219 0059

The Perm Contemporary Art Museum Competition

BERNASKONI
Tel: +7 495 697 8379
Valerio Olgiati
Zaha Hadid Architects
Tel: +44(0)20 7253 5147
Fax: +44(0)20 7251 8322
Acconci Studio
Guy Nordenson and Associates
Tel: 212 766 9119
Fax: 212 766 9016
Alexander Brodsky Bureau
Asymptote Architecture
Tel: 212 343 7333
Fax: 718 937 3320
Esa Ruskeepaa
Tel: +358414604106
AB Architects
Tel: (085) 74 32 381
Fax: (085) 74 32 319
Soren Robert Lund Arkitekter
Tel: +45 33910100
Fax: +45 33914510
Totan Kuzembaev Architectural Workshop
Tel: +7-495-236-45-73
Fax: +7-495-236-66-97

University of Michigan Museum Addition Competition

Allied Works Architecture
Tel: 503 227 1737
Fax: 503 227 6509
Polshek Partnership Architects
Tel: 212 807-7171
Fax: 212 807-5917
Weiss/Manfredi Architects
Tel: 212 760 9002
Fax: 212 760 9003

Whatcom Museum of History & Art and Whatcom Children's Museum Competition

Cambridge Seven Associates, Inc.
Tel: 617.492.7000
Fax: 617.492.7007

Czech National Library Competition

Future Systems
Carmody Groarke
Tel: +44 (0)207 8362333
Fax: +44 (0)207 8362334
HSH architekti
Emergent Architecture
Tel: (213) 385-1475
John Reed Architucture
Tel: 914 806 3724
MVMarchitekt
Tel: 0049-(0)221-8014583
Fax: 0049-(0)221-8014585
Holzer Kobler Architekturen
Tel: +41 44 240 52 00
Fax: +41 44 240 52 02

Kazakhstan Library Competition

BIG
Tel: +45 7221 7227

Milford Library Competition

Frederic Schwartz Architects
Tel: 212 741 3021
Bohlin Cywinski Jackson
Tel: 570.825.8756
Fax: 570.825.3744
BKSK Architects
Tel: 212 807 9600
Fax: 212 807 6405

Philadelphia's Public Library Competition

Moshe Safdie & Associates
Tel: 617 629 2100
Fax: 617 629 2406
TEN Arquitectos
Tel: 212 620 0794
Fax: 212 620 0798

Stockholm City Library Competition

Heike Hanada
Tel: + 49 (0) 30 - 31 01 86 60
Mauri Korkka
Tel: +358 9 622 1241
Paleko Archstudio
Tel: +49 (30) 44308505
Fax: +49 (30) 44308178